This report contains the collective views of an international group of experts and does not necessarily represent the decisions or the stated policy of the United Nations Environment Programme, the International Labour Organisation, or the World Health Organization.

Environmental Health Criteria 110

TRICRESYL PHOSPHATE

Published under the joint sponsorship of the United Nations Environment Programme, the International Labour Organisation, and the World Health Organization

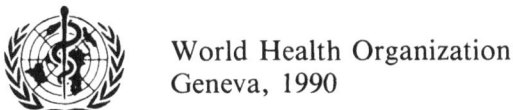

World Health Organization
Geneva, 1990

The **International Programme on Chemical Safety (IPCS)** is a joint venture of the United Nations Environment Programme, the International Labour Organisation, and the World Health Organization. The main objective of the IPCS is to carry out and disseminate evaluations of the effects of chemicals on human health and the quality of the environment. Supporting activities include the development of epidemiological, experimental laboratory, and risk-assessment methods that could produce internationally comparable results, and the development of manpower in the field of toxicology. Other activities carried out by the IPCS include the development of know-how for coping with chemical accidents, coordination of laboratory testing and epidemiological studies, and promotion of research on the mechanisms of the biological action of chemicals.

WHO Library Cataloguing in Publication Data

Tricresyl phosphate.

(Environmental health criteria ; 110)

1. Tritolyl phosphates - adverse effects   2. Tritolyl phosphates - toxicity   I. Series

ISBN 92 4 157110 1   (NLM Classification: QV 627)
ISSN 0250-863X

©World Health Organization 1990

Publications of the World Health Organization enjoy copyright protection in accordance with the provisions of Protocol 2 of the Universal Copyright Convention. For rights of reproduction or translation of WHO publications, in part or *in toto*, application should be made to the Office of Publications, World Health Organization, Geneva, Switzerland. The World Health Organization welcomes such applications.

The designations employed and the presentation of the material in this publication do not imply the expression of any opinion whatsoever on the part of the Secretariat of the World Health Organization concerning the legal status of any country, territory, city, or area or of its authorities, or concerning the delimitation of its frontiers or boundaries.

The mention of specific companies or of certain manufacturers' products does not imply that they are endorsed or recommended by the World Health Organization in preference to others of a similar nature that are not mentioned. Errors and omissions excepted, the names of proprietary products are distinguished by initial capital letters.

Printed in Finland
90/8588 — Vammala — 5000

CONTENTS

ENVIRONMENTAL HEALTH CRITERIA FOR
TRICRESYL PHOSPHATE

1. SUMMARY 11

    1.1 Identity, physical and chemical properties,
analytical methods 11
    1.2 Sources of human and environmental exposure 11
    1.3 Environmental transport, distribution, and
transformation 11
    1.4 Environmental levels and human exposure 12
    1.5 Effects on organisms in the environment 12
    1.6 Kinetics and metabolism 13
    1.7 Effects on experimental animals and
in vitro test systems 14
    1.8 Effects on humans 15

2. IDENTITY, PHYSICAL AND CHEMICAL PROPERTIES,
ANALYTICAL METHODS 16

    2.1 Identity 16
        2.1.1 Tricresyl phosphate 16
        2.1.2 Tri-*o*-cresyl phosphate 17
        2.1.3 Tri-*m*-cresyl phosphate 18
        2.1.4 Tri-*p*-cresyl phosphate 18
    2.2 Physical and chemical properties 18
    2.3 Conversion factor 20
    2.4 Analytical methods 21
        2.4.1 Extraction and concentration 21
        2.4.2 Clean-up procedures 24
        2.4.3 Gas chromatography and mass spectrometry 25
        2.4.4 Contamination of analytical reagents 25
        2.4.5 Other analytical methods 26

3. SOURCES OF HUMAN AND ENVIRONMENTAL
EXPOSURE 27

    3.1 Production levels and processes 27
        3.1.1 Accidental release 28
    3.2 Uses 29

4. ENVIRONMENTAL TRANSPORT, DISTRIBUTION,
AND TRANSFORMATION 33

    4.1 Transport and transformation in the environment 33

		4.1.1	Release to the environment	33
		4.1.2	Fate in water and sediment	34
		4.1.3	Biodegradation	35
		4.1.4	Water treatment	36
	4.2	Bioaccumulation and biomagnification		37
		4.2.1	Fish	37
		4.2.2	Plants	37

5. ENVIRONMENTAL LEVELS AND HUMAN EXPOSURE 39

    5.1 Environmental levels 39
        5.1.1 Air 39
        5.1.2 Water 42
        5.1.3 Soil 42
        5.1.4 Sediment 42
    5.2 General population exposure 43
        5.2.1 Drinking-water 43
        5.2.2 Fish 43
        5.2.3 Human tissues 44
    5.3 Occupational exposure 44

6. EFFECTS ON ORGANISMS IN THE ENVIRONMENT 45

    6.1 Unicellular algae 45
    6.2 Aquatic organisms 45
    6.3 Insects 49
    6.4 Plants 49

7. KINETICS AND METABOLISM 51

    7.1 Absorption 51
    7.2 Distribution 52
    7.3 Metabolic transformation 53
    7.4 Excretion 56

8. EFFECTS ON EXPERIMENTAL ANIMALS AND *IN VITRO* TEST SYSTEMS 58

    8.1 Single exposure 59
    8.2 Short-term exposure 60
    8.3 Skin and eye irritation 61
    8.4 Teratogenicity 61
    8.5 Reproduction 61
    8.6 Mutagenicity and carcinogenicity 64
    8.7 Neurotoxicity 64
        8.7.1 Experimental neuropathology 64

		8.7.2	Neurochemistry	66
		8.7.3	Interspecies sensitivity and variability to OPIDN	67
		8.7.4	Neurophysiology	69

9. EFFECTS ON HUMANS 70

 9.1 Historical background 70
 9.2 Occupational exposure 73
 9.3 Clinical features 73
 9.4 Prognosis 76
 9.5 Neurophysiological investigations 77
 9.6 Pathological investigations 77
 9.7 Laboratory investigations 78
 9.8 Treatment 79

10. EVALUATION OF HUMAN HEALTH RISKS AND EFFECTS ON THE ENVIRONMENT 80

 10.1 Evaluation of human health risks 80
   10.1.1 Exposure levels 80
   10.1.2 Toxic effects 81
 10.2 Evaluation of effects on the environment 81
   10.2.1 Exposure levels 82
   10.2.2 Toxic effects 82

11. RECOMMENDATIONS 83

REFERENCES 84

RESUME 101

EVALUATION DES RISQUES POUR LA SANTE HUMAINE ET DES EFFETS SUR L'ENVIRONNEMENT 107

RECOMMANDATIONS 111

RESUMEN 112

EVALUACION DE LOS RIESGOS PARA LA SALUD HUMANA Y DE LOS EFECTOS EN EL MEDIO AMBIENTE 118

RECOMENDACIONES 122

WHO TASK GROUP ON ENVIRONMENTAL HEALTH CRITERIA FOR TRICRESYL PHOSPHATE

Members

Dr S. Dobson, Institute of Terrestrial Ecology, Monks Wood Experimental Station, Abbots Ripton, Huntingdon, Cambridgeshire, England (*Chairman*)

Dr S. Fairhurst, Medical Division, Health and Safety Executive, Bootle, Merseyside, England *(Joint Rapporteur)*

Ms N. Kanoh, Division of Information on Chemical Safety, National Institute of Hygienic Sciences, Setagaya-ku, Tokyo, Japan

Dr A. Nakamura, Division of Medical Devices, National Institute of Hygienic Sciences, Setagaya-ku, Tokyo, Japan

Dr M. Tasheva, Department of Toxicology, Institute of Hygiene and Occupational Health, Sofia, Bulgaria

Dr B. Veronesi, Neurotoxicology Division, US Environmental Protection Agency, Research Triangle Park, North Carolina, USA

Mr W.D. Wagner, Division of Standards Development and Technology Transfer, National Institute for Occupational Safety and Health, Cincinnati, Ohio, USA

Dr R. Wallentowicz, Exposure Assessment Application Branch, US Environmental Protection Agency, Washington, DC, USA *(Joint Rapporteur)*

Dr Shen-Zhi Zhang, Beijing Municipal Centre for Hygiene and Epidemic Control, Beijing, China

Observers

Dr M. Beth, Berufsgenossenschaft der Chemischen Industrie (BG Chemie), Heidelberg, Federal Republic of Germany

Dr R. Kleinstück, Bayer AG, Leverkusen, Federal Republic of Germany

Secretariat

Dr M. Gilbert, International Programme on Chemical Safety, Division of Environmental Health, World Health Organization, Switzerland (*Secretary*)

NOTE TO READERS OF THE CRITERIA DOCUMENTS

Every effort has been made to present information in the criteria documents as accurately as possible without unduly delaying their publication. In the interest of all users of the environmental health criteria documents, readers are kindly requested to communicate any errors that may have occurred to the Manager of the International Programme on Chemical Safety, World Health Organization, Geneva, Switzerland, in order that they may be included in corrigenda, which will appear in subsequent volumes.

* * *

A detailed data profile and a legal file can be obtained from the International Register of Potentially Toxic Chemicals, Palais des Nations, 1211 Geneva 10, Switzerland (Telephone No. 7988400 or 7985850).

ENVIRONMENTAL HEALTH CRITERIA FOR TRICRESYL PHOSPHATE

A WHO Task Group meeting on Environmental Health Criteria for Tricresyl Phosphate was held at the British Industrial Biological Research Association (BIBRA), Carshalton, United Kingdom, from 9 to 13 October 1989. Dr S.D. Gangolli, Director, BIBRA, welcomed the participants on behalf of the host institution and Dr M. Gilbert opened the meeting on behalf of the three cooperating organizations of the IPCS (ILO, UNEP, WHO). The Task Group reviewed and revised the draft criteria document and made an evaluation of the risks for human health and the environment from exposure to tricresyl phosphate.

The first draft of this document was prepared by DR A. NAKAMURA, National Institute for Hygienic Sciences, Japan. Dr M. Gilbert and Dr P.G. Jenkins, both members of the IPCS Central Unit, were responsible for the overall scientific content and editing, respectively.

ABBREVIATIONS

ACh	acetylcholine
AChE	acetylcholinesterase
BCF	bioconcentration factor
CNS	central nervous system
FPD	flame photometric detector
GC	gas chromatography
GLC	gas liquid chromatography
GPC	gel permeation chromatography
IC_{50}	inhibition concentration, median
LC_{50}	lethal concentration, median
MS	mass spectrometry
NOEL	no-observed-effect level
NPD	nitrogen-phosphorus sensitive detector
NTE	neurotoxic esterase
OPIDN	organophosphate-induced delayed neuropathy
2-PAM chloride	pralidoxine (2-pyridine aldoxime methyl) chloride
PVC	polyvinyl chloride
TAP	triaryl phosphate
TBP	tributyl phosphate
TCP	tricresyl phosphate
TLC	thin-layer chromatography
TMCP	tri-*m*-cresyl phosphate
TOCP	tri-*o*-cresyl phosphate
TPCP	tri-*p*-cresyl phosphate
TPP	triphenyl phosphate

1. SUMMARY

1.1 Identity, physical and chemical properties, analytical methods

Tricresyl phosphate (TCP) is a non-flammable, non-explosive, colourless, viscous liquid. Its partition coefficient between octanol and water (log P_{ow}) is 5.1. It is easily hydrolysed in an alkaline medium to produce dicresyl phosphate and cresol, but it is stable in neutral and acidic media at normal temperatures.

The analytical method of choice is gas chromatography with a nitrogen-phosphorus sensitive detector or a flame photometric detector. The detection limit in a water sample is approximately 1 ng/litre. TCP is easily extracted from aqueous solution with various organic solvents. Florisil column chromatography is usually used for clean-up, but it is difficult to separate TCP from lipids by this method. Other clean-up methods (GPC, activated charcoal chromatography and Sep-pak C-18) have been recommended for the purpose. Analytical reagents are often contaminated with traces of TCP because of its widespread use. Therefore, care must be taken in order to obtain reliable data in trace analysis of TCP.

1.2 Sources of human and environmental exposure

TCP is usually produced by the reaction of cresols with phosphorus oxychloride. There are two industrial sources of cresols: "cresylic acid" obtained as a residue from coke ovens and petroleum refining; and "synthetic cresols" prepared from cymene via oxidation and degradation. As a result, TCP is a mixture of various triaryl phosphates.

TCP is used as a plasticizer in vinyl plastics, as a flame-retardant, as an additive to extreme pressure lubricants, and as a non-flammable fluid in hydraulic systems.

1.3 Environmental transport, distribution, and transformation

The release of TCP to the environment derives mainly from end-point use, little release occurring during

Summary

production. The total release to the environment in the USA was estimated at 32 800 tonnes in 1977.

Because of its low water solubility and high adsorption to particulates, TCP is rapidly adsorbed onto river or lake sediment and soil. Its biodegradation in the aquatic environment is rapid, being almost complete in river water within 5 days. The ortho isomer is degraded slightly faster than the meta or para isomers. TCP is readily biodegraded in sewage sludge with a half-life of 7.5 h, the degradation within 24 h being up to 99%. Abiotic degradation is slower with a half-life of 96 days.

Bioconcentration factors (BCF) of 165-2768 were measured for several fish species in the laboratory using radiolabelled TCP. The radioactivity was lost rapidly on cessation of exposure, depuration half-lives ranging between 25.8 and 90 h.

1.4 Environmental levels and human exposure

TCP has been measured in air at concentrations up to 70 ng/m^3 in Japan but reached a maximum of only 2 ng/m^3 at a production site in the USA. Workplace air in the USA contained less than 0.8 mg/m^3 at a lubrication oil barrel-filling operation and 0.15 mg/m^3 (total phosphates) in an automobile zinc die-casting plant. Concentrations of TCP measured in drinking-water in Canada were low (0.4 to 4.3 ng/litre) and TCP was undetectable in well-water. Levels in river and lake waters are frequently considerably higher. However, this is due to the presence of suspended sediment to which TCP is strongly adsorbed.

Concentrations in sediment are higher with up to 1300 ng/g in river sediment and 2160 ng/g in marine sediment.

Levels in soil and vegetation measured within the perimeter of production plants were elevated.

Residues in fish and shellfish of up to 40 ng/g have been reported but the majority of sampled animals contained no detectable residues.

1.5 Effects on organisms in the environment

The primary productivity of cultures of freshwater green algae was reduced to 50% by tri-*o*-cresyl phosphate

(TOCP) at 1.5 to 4.2 ng/litre, depending on the species, whereas the meta and para isomers were less toxic. There are limited data on the acute toxicity of TCP to aquatic invertebrates: the 48-h LC_{50} for Daphnia is 5.6 ng/litre; the 24-h LC_{50} for nematodes is 400 ng/litre; the 2-week NOEL for Daphnia (mortality, growth, reproduction) is 0.1 mg/litre. The 96-h LC_{50} values for three fish species were between 4.0 and 8700 mg/litre. Rainbow trout showed approximately 30% mortality after a 4-month exposure to 0.9 ng/litre IMOL S-140 (2% tri-*o*-cresyl phosphate, TOCP) and minor effects within 14 days.

The exposure levels used in these experiments were much greater than likely water concentrations in the environment and, in most cases, greatly exceed the solubility of the compounds.

1.6 Kinetics and metabolism

The absorption, distribution, metabolism, and elimination of organophosphates are critical to the delayed neuropathic effects of these compounds.

Dermal absorption of TOCP in humans appears to be at least an order of magnitude faster than that in dogs. Significant dermal absorption also appears to occur in cats. Oral absorption of the compound has been reported in rabbits. There is no direct information on absorption via the inhalation route.

In cat studies, absorbed TOCP was widely distributed throughout the body, the highest concentration being found in the sciatic nerve, a target tissue. Other tissues with high concentration of TOCP and its metabolites were the liver, kidney, and gall bladder.

TOCP is metabolized via three pathways. The first is the hydroxylation of one or more of the methyl groups, and the second is dearylation of the *o*-cresyl groups. The third is further oxidation of the hydroxymethyl to aldehyde and carboxylic acid. The hydroxylation step is critical because the hydroxymethyl TOCP is cyclized to form saligenin cyclic *o*-tolyl phosphate, the relatively unstable neurotoxic metabolite.

Summary

TOCP and its metabolites are eliminated via the urine and faeces, together with small amounts in the expired air.

1.7 Effects on experimental animals and *in vitro* test systems

Of the three isomers of TCP, TOCP is by far the most toxic in acute and short-term exposure. It is the only isomer that produces delayed neurotoxicity.

There is wide interspecies variability for the various toxic end-points (e.g., acute lethality, delayed neurotoxicity) of TOCP exposure, the chicken being one of the most sensitive species.

Organophosphate-induced delayed neuropathy (OPIDN) has been produced with both single and repeated exposure regimes in a wide range of experimental species and it is classified as a "dying-back neuropathy". Degenerative changes occur in the distal axon and extends with time towards the cell body.

Clinical signs are paralysis of the hindlegs after a characteristic delay of 2-3 weeks after exposure. A single oral dose of 50-500 mg TOCP/kg induced delayed neuropathy in chickens, whereas doses of 840 mg/kg or more were necessary to produce spinal cord degeneration in Long-Evans rats. The metabolite saligenin cyclic *o*-tolyl phosphate is the active neurotoxic agents. Species sensitivity is inversely correlated with rate of further metabolism.

Inhibition of "neurotoxic esterase" is thought to be the biochemical lesion leading to OPIDN; inhibition by more than 65% shortly after exposure to TOCP presages subsequent neuropathy. Factors other than metabolism (e.g., route of exposure, age, sex, strain) influence variability in response to TOCP neurotoxicity. A clear no-observed-effect level for delayed neuropathy is not apparent from the data available.

Reproduction studies in rats and mice receiving repeated oral exposure to TOCP showed histopathological damage in the testes and ovaries, morphological changes in sperm, decreased fertility in both sexes, and decreased litter size and viability. A clear no-effect level for the reproductive effects of TOCP was not apparent from the data available. A teratogenicity study in rats, using oral doses producing maternal toxicity, yielded negative results.

Little information is available on mutagenicity and none on carcinogenicity.

1.8 Effects on humans

Accidental ingestion is the main cause of intoxication. Since the end of the nineteenth century, numerous cases of poisoning due to contamination of drink, food, or drugs have been reported. Occupational exposure is principally via dermal absorption or inhalation, and some cases of poisoning have been reported. Ingestion of preparations contaminated by TOCP may be followed by gastrointestinal symptoms (nausea, vomiting, and diarrhoea), although in some cases polyneuropathy is the first evidence of poisoning. The neurological symptoms are characteristically delayed. The initial symptoms are pain and paraesthesia in the lower extremities. A mild impairment of cutaneous sensations and sometimes an impairment of vibratory sense may be present. In most cases the muscle weakness progresses rapidly to a striking paralysis of the lower extremities with or without an involvement of the upper extremities. Severe cases show pyramidal signs. Fatalities are rare, but recovery from the neurological signs and symptoms can be extremely slow and extend over a number of months or years. Histopathological findings show axonal degeneration. Routine laboratory examinations show no abnormal findings, but an increase of protein concentration in the cerebrospinal fluid may be seen. First aid should reduce exposure by inducing vomiting immediately after ingestion, providing the patient is conscious. The cardinal long-term therapy is physical rehabilitation and no specific antidote is known. There is considerable variation between individuals both in response to TCP and recovery from the toxic effects. Severe symptoms have been reported following the ingestion of 0.15 g of TCP, while other individuals failed to show any toxic effect after ingesting 1-2 g. Some patients show complete recovery, whereas others retain marked effects for a considerable period.

2. IDENTITY, PHYSICAL AND CHEMICAL PROPERTIES, ANALYTICAL METHODS

2.1 Identity

2.1.1 Tricresyl phosphate (commercial product: mixture of isomers)

Chemical structure:

Molecular formula:	$C_{21}H_{21}O_4P$
Relative molecular mass:	368.4
CAS chemical name:	phosphoric acid, tritolyl ester
CAS registry number:	1330-78-5
RTECS registry number:	TD0175000
Synonyms:	tricresylphosphate, tricresyl phosphate, TCP, tritolyl phosphate, trimethylphenyl phosphate
Trade name:	Kronitex-TCP®, Santicizer 140®, Pliabrac 521®, Phosflex 179®, Disflamoll TKP®, Lindol®,

Kolflex 50⁵⁰®, PX.917®,
Celluflex 179C®,

Manufacturers and suppliers (Modern Plastics Encyclopedia, 1975):

Albright & Wilson Ltd., Ashland Chemical Co., Bayer AG, Celanese Co., East Coast Chemicals Co., F.M.C. Corp., Harwick Chemical Corp., Kolker Chemical Co., McKesson Chemical Co., Mobay Chemical Co., Pittsburgh Chemical Co., Rhone-Poulenc Co., Sobin Chemical Co., Stauffer Chemical Co., Daihachi Chemical Ind. Co. Ltd., Kyowa Hakko Kogyo Co. Ltd., Hodogaya Chemical Co. Ltd., Mitsubishi Gas Chemical Co. Inc., Kurogane Kasei Co. Ltd., Kashima Ind. Co.

2.1.2 Tri-o-cresyl phosphate

Chemical structure:

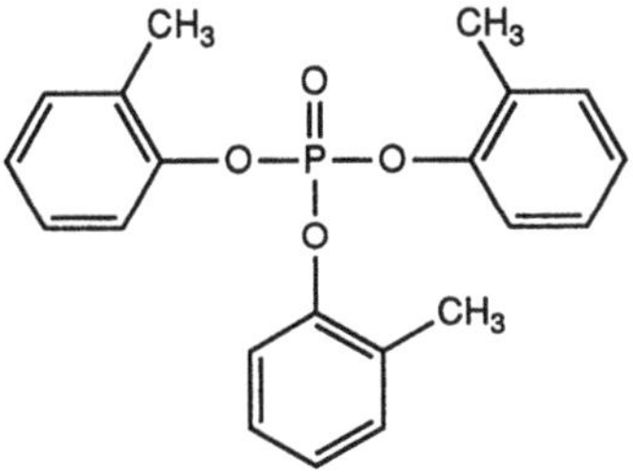

CAS chemical name: phosphoric acid, tri-o-tolyl ester

CAS registry number: 78-30-8

RTECS registry
number: TD0350000

Synonyms: tri-o-cresyl phosphate, tri-o-cresylphosphate, phosphoric acid tris(2-methylphenyl) ester, o-TCP, TOCP, TOTP, tri-o-tolyl phosphate, tri-2-tolyl phosphate, tri-2-methylphenyl phosphate

2.1.3 Tri-m-cresyl phosphate

Chemical structure:

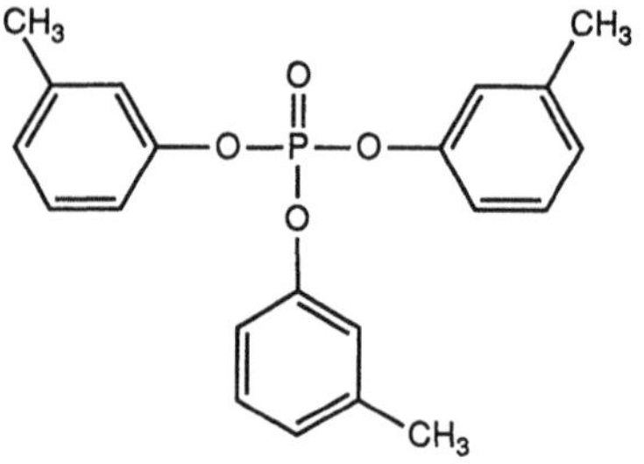

CAS chemical name: phosphoric acid, tri-*m*-tolyl ester
CAS registry number: 563-04-2
Synonyms: tri-*m*-cresylphosphate, phosphoric acid tris(3-methylphenyl) ester, *m*-TCP, tri-*m*-tolyl phosphate, tri-3-tolyl phosphate, tri-3-methylphenyl phosphate

2.1.4 Tri-p-cresyl phosphate

Chemical structure:

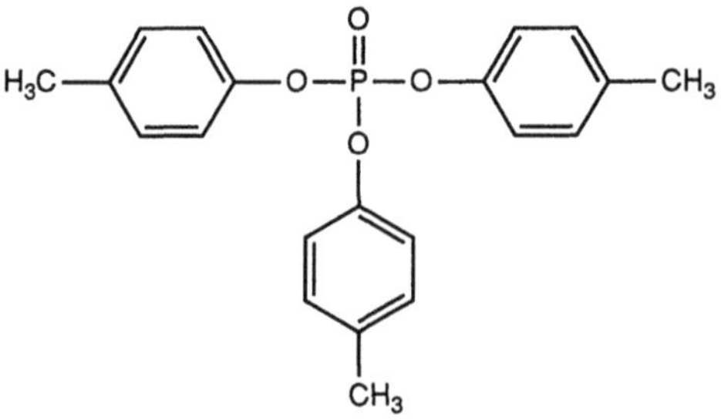

CAS chemical name: phosphoric acid, tri-*p*-tolyl ester
CAS registry number: 78-32-0
Synonyms: tri-*p*-cresylphosphate, phosphoric acid tris(4-methylphenyl) ester, *p*-TCP, tri-*p*-tolyl phosphate, tri-4-tolyl phosphate, tri-4-methylphenyl phosphate

2.2 Physical and chemical properties

The physical properties of tricresyl phosphate (TCP) are listed in Table 1.

Table 1. Physical properties of tricresyl phosphate and isomers

Physical properties	Tricresyl phosphate (mixtures of isomers)	Tri-o-cresyl phosphate	Tri-m-cresyl phosphate	Tri-p-cresyl phosphate
Physical state	liquid	liquid	half-solid	crystalline solid
Colour	colourless	colourless	colourless	colourless
Odour	very slightly aromatic	very slightly aromatic	very slightly aromatic	very slightly aromatic
Melting or freezing point (°C)	-33[b]	11[a]	25.6[a]	77-78[a]
Boiling point or range (°C)	241-255 (4 mmHg)[b]; 190-200 (0.5-10 mmHg)[c]	410 (760 mmHg)[a]	260 (15 mmHg)[a]	244 (3.5 mmHg)[a]
Specific gravity (density)	1.160-1.175 (25 °C)[b]; 1.165[c]	1.1955[a]	1.150[a]	1.237[a]
Refractive index	1.553-1.556 (25 °C)[b]; 1.556 (20 °)[c]	1.5575[a]	1.5575[a]	
Viscosity (cSt)	60 (25 °C), 4.0 (100 °C)[c]			
Flash point (°C)	257[c]			
Vapour pressure (mmHg)	1×10^{-4} (20 °C)[c]	10 (265 °C)[a]		
Henry's Law constant	1.1-2.8×10^{-6} atm-m^3/mol[d]			
Solubility in water (mg/litre)	0.36[e]; 0.34 ± 0.04[f]			0.074[g]
Octanol-water partition coefficient (log P_{ow})	5.11[e]; 5.12[h]			

[a] Hine et al. (1981).
[b] Modern Plastics Encyclopedia (1975).
[c] Lefaux (1972).
[d] Boethling & Cooper (1985).
[e] Saeger et al. (1979).
[f] Ofstad & Sletten (1985).
[g] Hollifield (1979).
[h] Kenmotsu (1980b).

TCP is non-flammable and non-explosive. When the para isomer was heated at 370 °C with air for 30 min, 99% of the compound was recovered. The main volatile products obtained were water, carbon dioxide, toluene, and cresols (Paciorek et al., 1978). No data on pyrolysis or combustion of TCP at higher temperatures are available (at about 600 °C, triphenyl phosphate begins to decompose, yielding some aromatic hydrocarbons, some oxygenated aromatic compounds, and phosphoric oxides). Its partition coefficient between octanol and water (log P_{ow}) is approximately 5.11-5.12.

Hydrolysis of TCP is thought to proceed in an analogous manner to triphenyl phosphate. It hydrolyses rapidly in an alkaline solution. Despite a lack of data, neutral or acidic hydrolysis of TCP, by analogy to TPP, is assumed to be very slow. The hydrolysis rate constants and half-lives reported are summarized in Table 2. Formation of dicresyl phosphate during alkaline hydrolysis would be expected, but no data are available (Wolfe, 1980).

Table 2. Hydrolysis rate constant (2nd order, K_2) and half-lives in aqueous solution

Compound	Solution	Temperature (°C)	pH	Rate constant ($M^{-1}.sec^{-1}$)	Half-life	Reference
Tri-*p*-cresyl phosphate	Water 0.2N NaOH/ acetone (1 : 1)	27 22	alkaline 13	2.5 x 10^{-1}	 1.66 h	Wolfe (1980) Muir et al. (1983)
Tri-*m*-cresyl phosphate	0.1N NaOH/ acetone (1 : 1)	22	13		1.31 h	Muir et al. (1983)

The photolysis of TAPs in ethanol yielded the corresponding monoaryl phosphate and diphenyl derivatives (Finnegan, 1972). The results are summarized in Table 3.

2.3 Conversion factor

Tricresyl phosphate   1 ppm = 15.07 mg/m³ air

Table 3. Photolysis of symmetrical triaryl phosphates[a]

Starting compound	Resulting compounds		Recovered ester (%)	Quantum yield for biaryl formation
	Ar-Ar (%)	ArOPO$_3$H$_2$ (%)		
Phenyl	2		48	6 × 10^{-4}
p-Tolyl	35-51	2-10	13-20	190 × 10^{-4}
p-t-butylphenyl	51	55	24	44 × 10^{-4}
Mesityl	4	7	7	not determined

[a] From: Finnegan & Matson (1972).
The esters were irradiated, at a concentration of 0.02 mol/litre ethanol, using a 450W Hanovia arc lamp, for 5 h.

2.4 Analytical methods

Analytical methods for determining TCP in air, water, sediment, fish, biological tissues, and edible oils are summarized in Table 4. The method of choice is gas chromatography (GC) with a nitrogen-phosphorus sensitive detector (GC/NPD) or a flame photometric detector (GC/FPD). The detection limit in water samples is at the ng/litre level. Using GC, TCP and other trialkyl/aryl phosphates, such as triphenyl phosphate (TPP), trioctyl phosphate, and trixylenyl phosphate, can be simultaneously determined. High-performance liquid chromatography (HPLC) and thin-layer chromatography (TLC) are sometimes used for determining TCP, but these are not widely applicable.

It should be noted that the behaviour of TCP in analytical processes and in its environmental distribution is similar to that of other TAPs, lipids, and phthalic acid esters, owing to analogous physical and chemical properties.

2.4.1 Extraction and concentration

TCP is easily extracted from aqueous solution with methylene chloride, hexane, or benzene (Kenmotsu et al., 1980a; Muir et al., 1981). Low levels of TCP in water can be successfully concentrated on an Amberlite XAD-2 resin

Table 4. Methods for the determination of TCP and TPP

Sample type	Sampling method extraction/clean-up	Analytical method	Limit of detection	Applicability	Reference
Workplace air	collect with Millipore filter, extract with ethanol	GC/FPD	1 µg per sample	TCP and TPP	US NIOSH (1982)
Environmental air	trap with glycerol-Florisil column, eluate with methanol, add water, and extract with hexane	GC/FPD	1 ng/m^3	simultaneous method for trialkyl/aryl phosphates	Yasuda (1980)
Air	collect by aspiration through ethanol, hydrolyse with NaOH; the resultant phenols are reacted with $p\text{-}O_2NC_6H_4N_2^+$ and separated with silica gel plate	TLC	5 ng/plate	TCP and TPP	Druyan (1975)
Drinking-water	adsorb with XAD-2 resin, eluate with acetone-hexane or acetone	GC/NPD GC/MS	1 ng/litre	method for low level trialkyl/aryl phosphates	Lebel et al. (1979, 1981)
River or sea water	extract with methylene chloride or benzene	GC/NPD GC/FPD GC/MS	0.02 µg/litre (TPP) 0.05 µg/litre (TCP)	simultaneous method for trialkyl/aryl phosphates	Kenmotsu et al. (1980a, 1981b, 1982b) Muir et al. (1981) Ishikawa et al. (1985)
Farm pond sediment	reflux with methanol-water (9+1) or methylene chloride-methanol (1+1), clean-up by acid alumina column chromatography	GC/NPD	1 ng/g	simultaneous method for triaryl phosphates	Muir et al. (1980, 1981)
River or sea sediment	extract with acetonitrile or acetone, clean-up by charcoal or Florisil column chromatography	GC/FPD GC/MS	5 ng/g	simultaneous method for trialkyl/aryl phosphates	Kenmotsu et al. (1980a, 1981b, 1982a, 1982b, 1983) Ishikawa et al. (1985)

Table 4 (contd).

Sample type	Sampling method extraction/clean-up	Analytical method	Limit of detection	Applicability	Reference
Fish	extract with hexane or methanol, clean-up by gel permeation column chromatography and acid alumina column chromatograpy	GC/NPD GC/MS	1 ng/g	simultaneous method for triaryl phosphates	Muir et al. (1980, 1981, 1983)
Fish	extract with acetonitrile and methylene chloride, clean-up by acetonitrile-hexane partitioning, charcoal column chromatography, concentrated sulfuric acid extraction and Florisil column chromatography	GC/FPD GC/MS	5 ng/g	simultaneous method for trialkyl/aryl phosphates	Kenmotsu et al. (1980a)
Human adipose tissues	extract with benzene or acetone-hexane (15 + 85), clean-up by gel permeation chromatography and Florisil column chromatography	GC/NPD GC/FPD GC/MS	1 ng/g	simultaneous method for trialkyl/aryl phosphates	Lebel & Williams (1983)
Plasma	extract with ethyl ether, filter with 0.45-μm nylon filter	HPLC (254 nm)	50 ng/injection	TCP and its metabolites	Nomeir & Abou-Donia (1983)
Edible oils	extract with ethanol, hydrolyse with NaOH; the resultant cresol is coupled with 2,6-dichlorobenzoquinone	colorimetric	0.01%	simple method for TCP determination	Vaswani et al. (1983)
Edible oils	separate with silica gel G thin-layer plate; spray rhodamine B solution	TLC (UV)		simple method for TCP determination	Bhattacharyya et al. (1974)

column (Lebel et al., 1981; Lebel & Williams, 1983). TCP has been extracted from sediment with various polar solvents, such as aqueous methanol (Muir et al., 1980, 1981), acetonitrile (Kenmotsu et al., 1980a), or acetone (Ishikawa et al., 1985). The extraction method established by the US Association of Official Analytical Chemists (AOAC) for organochlorine and organophosphorus pesticides is also applicable for the extraction of TCP from fat-containing foods and fish (Lombardo & Egry, 1979). Hexane (Lombardo & Egry, 1979), methanol (Muir et al., 1980, 1981; Muir & Grift, 1983), acetonitrile and methylene chloride (Kenmotsu et al., 1979), and acetone-hexane (Lebel & Williams, 1983) have been used for the extraction of TCP from fish or adipose tissue. Workplace airborne samples can be collected on Millipore® filters and the particulate TCP analysed (US NIOSH, 1977, 1979, 1982). Vapour phase and particulate TCP in the atmosphere have been simultaneously collected on glycerol-coated Florisil® columns and 96% of the TCP recovered (Yasuda, 1980). The Midwest Research Institute (MRI/USA) has used high-volume air filter pads and activated carbon filters to sample ambient air (MRI, 1979).

2.4.2 Clean-up procedures

Florisil column chromatography has been used routinely for clean-up of TCP (Lombardo & Egry, 1979; Kenmotsu et al., 1980a; Lebel & Williams, 1983). The separation of TCP from tributyl phosphate (TBP) and parathion is possible by this procedure but is more difficult than for other TAPs such as trixylenyl phosphate (Kenmotsu et al., 1981b; Lebel & Williams, 1983). Sulfur-containing compounds, which often exist in sediment samples and interfere with the analysis of TCP by GC/FPD, can easily be separated by elution with hexane from Florisil columns (Kenmotsu et al., 1980a). Partitioning between acetonitrile and petroleum ether is useful to separate TCP from fish fat (Lombardo & Egry, 1979; Kenmotsu et al., 1980a). Since the polarity of TCP is similar to that of lipids in biological tissues, it is difficult to separate TCP from lipids by Florisil column chromatography. Gel permeation chromatography (GPC) is useful in this case (Muir et al., 1981), the elution volume varying according to the type of phosphate ester, i.e. trialkyl-, triaryl-, or tri(haloalkyl) phosphates (Lebel & Williams, 1983). Activated charcoal column chromatography (Kenmotsu et al., 1980a), alumina column chromatography (Muir et al., 1980, 1981), and C-18 bonded silica cartridge (Sep-pak C-18) (Muir et al., 1980; Muir & Grift, 1983) have also been used to separate TCP from co-extracting compounds in various samples.

2.4.3 Gas chromatography and mass spectrometry

To identify TCP in environmental samples by packed column GLC, it is useful to compare retention times using two types of liquid phase with different polarities. As a low polarity liquid phase, 10% OV-1 (Kenmotsu et al., 1980a), 3% SE-30 (Ramsey & Lee, 1980), 3% OV-17 (Lebel et al., 1981), 3% OV-101 (Deo & Howard, 1978), SP-2100 (Muir et al., 1980), and 5% DC-200 (Daft, 1982) have been used, while 1% QF-1 (Bloom, 1973), 5% FFAP and 5% Thermon-3000 (Kenmotsu et al., 1980a), and 2% DEGS (Daft, 1982) have been used as a higher polarity liquid phase.

TCP is often accompanied by other TAPs in environmental samples, which show multiple peaks in GC and occasionally have the same retention indices as that of TCP (Ramsey & Lee, 1980; Kenmotsu et al., 1982b). Therefore, capillary GLC or GC-mass spectrometry (GC/MS) is preferred (Lebel et al., 1981; Lebel & Williams, 1983; Kenmotsu et al., 1983; Ofstad & Sletten, 1985). In electron impact mass spectrometry, TCP gives a high intensity molecular ion, as do other TAPs (Deo & Howard, 1978; Wightman & Malaiyandi, 1983; Kenmotsu et al., 1982b). A selected ion monitoring (SIM) technique is also useful for trace analysis of TCP in environmental samples (Ishikawa et al., 1985), but care must be taken to select suitable fragment ions in order to avoid interference by other TAPs.

The phenolic components of TCP are confirmed by alkaline hydrolysis, followed by GLC analysis of the resulting phenols (Murray, 1975; Sugden et al., 1980).

2.4.4 Contamination of analytical reagents

The widespread use of TCP in plastics and hydraulic fluids can cause contamination of analytical reagents. Traces of TCP have been found in rubber O-rings and rubber seals used in a Corning water supply system (Lebel et al., 1981), Super Q water (Williams & Lebel, 1981), and acetonitrile, methylene chloride, and hexane (Daft, 1982). Trialkyl phosphates have also been found in cyclohexane (Bowers et al., 1981), hexane (Hudec et al., 1981), and analytical grade filters (Daft, 1982). Therefore, care must be taken to avoid contamination of analytical reagents in order to obtain accurate data in trace analysis of TCP.

2.4.5 Other analytical methods

A rapid colorimetric method has been developed for the determination of TCP in edible oil (Vaswani et al., 1983), but no information about the interference with other TAPs is available. Silica gel TLC has been used for determining TCP in edible oil (Bhattacharyya et al., 1974; Krishnamurthy et al., 1985). Reversed phase TLC has also been used (Renberg et al., 1980). However, separating TAPs from each other by TLC is not sufficient (Bloom, 1973). HPLC with a C-18 bonded column has been used for determining TCP in plasma, while size exclusion HPLC has been used in the case of machine oil (Majors & Johnson, 1978). An ultraviolet spectrometric detector is usually used in HPLC, but it is not specific for TAPs. Tittarelli & Mascherpa (1981) described a highly specific HPLC detector for TAPs using a graphite furnace atomic absorption spectrometer.

3. SOURCES OF HUMAN AND ENVIRONMENTAL EXPOSURE

3.1 Production levels and processes

Tricresyl phosphate does not occur naturally in the environment. Figures concerning the total world production are not available. In Japan, 33 000 tonnes were produced in 1984[a]. In the USA, approximately 54 000 tonnes of TAP including 10 400 tonnes of TCP were produced in 1977 (Boethling & Cooper, 1985). About 800-1000 tonnes TCP per year is now produced in China.

TCP is usually produced by the reaction of cresols with phosphorus oxychloride. One of the industrial sources of cresols is the so-called cresylic acid or tar acid, which is a mixture of isomers of cresol and varying amounts of xylenols, phenol, and other high-boiling phenolic fractions obtained as a residue from coke ovens and petroleum refining (Duke, 1978). Another source is "synthetic cresol", prepared from cymene via oxidation and catalytic degradation (Association of the Plasticizer Industry of Japan, 1976), and this has been used for production of TCP in Japan since 1971. The composition of some cresylic acids and synthetic cresol is shown in Table 5.

TCP derived from these alkylphenols is, therefore, a complex mixture of various TAPs, i.e. tri-*o*-cresyl phosphate, tri-*m*-cresyl phosphate, tri-*p*-cresyl phosphate, di-*m*-cresyl-*p*-cresyl phosphate, di-*p*-cresyl-*m*-cresyl phosphate, etc. The very toxic tri-*o*-cresyl phosphate (TOCP) is usually excluded as much as possible. In some cases, commercial tricresyl phosphate (TCP) has been reported to contain a small amount of TPP (Daft, 1982; Ofstad & Sletten, 1985).

The most noteworthy trend in aryl phosphate manufacture and use in the USA has been the replacement of triphenyl, tricresyl, and trixylenyl phosphates derived from

---

[a] Personal communication from the Association of the Plasticizer Industry of Japan, 1985.

Table 5. Composition of some commercial cresylic acids[a] and "synthetic cresol"[b]

Constituents	Boiling point (°C)	Composition (%) Cresylic Acids			Synthetic Cresol
		Sample A	Sample B	Sample C	
o-Cresol	191.0	3	0	0	0.1%
2,6-Xylenol	201.0	6	6	0	
m-Cresol	202.2	42	43	47	} 99%
p-Cresol	202.3	30	31	34	
o-Ethylphenol	204.5	3	3	0	
2,4-/2,5-Xylenol	211.5	16	17	19	

[a] From: Bondy et al. (1960).
[b] From: Association of the Plasticizer Industry of Japan (1976).

petroleum-based feedstocks by aryl phosphates derived from synthetic precursors. The production of cresyl diphenyl phosphate, a petroleum-based aryl phosphate, was discontinued in 1979 in the USA. The mixed tri-alkyl/aryl phosphates are replacing TPP and TCP as a plasticizer, whereas the synthetic TAPs are replacing TCP and trixylenyl phosphate in functional fluids (Boethling & Cooper, 1985).

3.1.1 Accidental release

Liquid TCP and hydraulic fluid and lubricant oil containing phosphate esters are transported by tank trucks, rail cars, and to a lesser extent in barrels (US NIOSH, 1979). Occasionally, empty barrels (or drums) previously containing hydraulic fluid or lubricant oil have been reused to store or to transport edible oil (or water), and this has resulted in poisoning of humans and cattle (Susser & Stein, 1957; Smith & Spalding, 1959; Chaudhuri, 1965; Nicholson, 1974; Senanayake, 1981). Another case of poisoning involved flour contaminated with oil from a leak during shipping (Sorokin, 1969).

In a report by Beck et al. (1977), accidental spillage of TAPs intended for use in pipeline pumping stations occurred and resulted in poisoning of cattle. Effluent from an evaporation pond overflowed onto the pasture during spring run-off. Sampling showed concentrations of TAPs from 0.304% to 3.44% by weight in soil, grass, and water near the plant. Thirty days later TAPs were still present in the evaporation pond but not in the soil samples (Chemical and Geological Laboratories Ltd., 1971).

Beck et al. (1977) described in his report: "Mass poisonings are possible because large quantities of triaryl phosphate are used as lubricants and coolants in jet engines in pipeline compressor stations. If an emergency arises as much as 1200 gallons of this material can be expelled into the atmosphere within 20 seconds. The construction of pipelines over thousands of miles, with manned or unmanned compressors every 100 miles, constitutes an environmental hazard to both domestic livestock and wildlife."; and "Natural leaching of the ground by weather conditions probably removed the poison from the soil, but repeated spills of large quantities of a stable compound could contaminate ground water supplies".

3.2 Uses

TCP is used as a plasticizer in vinyl plastic manufacture, as a flame-retardant, a solvent for nitrocellulose, in cellulosic molding compositions, as an additive to extreme pressure lubricants, and as a non-flammable fluid in hydraulic systems (Windholz, 1983). The main market for PVC-based products plasticized with organic phosphate esters is in the manufacture of automobile and other motor vehicle interiors in the USA (Lapp, 1976). In Japan, approximately 2500 tonnes of TCP was used in 1984 as a plasticizer in PVC film for agricultural use, 400 tonnes as non-flammable plasticizer in floor and wall covering, and 100 tonnes for miscellaneous purposes (Association of the Plasticizer Industry of Japan, 1985).

The fastest growing use of organic phosphate esters is in the manufacture of fire-resistant hydraulic fluids and lubricants in the USA (Lapp, 1976). The two types of organic phosphate hydraulic fluids being manufactured are phosphate ester oil blends and "pure synthetics". The phosphate ester oil blends contain between 30% and 50% organic phosphate esters in addition to petroleum oil and coupling agents; the "pure synthetics" contain a mixture of organic phosphate esters. For example, a typical synthetic organic phosphate fluid contains TCP, trixylenyl phosphate, and other TAPs. The compositions of several commercial synthetic organic phosphate fluids are listed in Table 6. Organic phosphate ester lubricant additives are usually of three general types: extreme pressure agents, anti-wear agents, and stick-slip moderators. The first two types are used in systems with some type of gears and account for over 80% of all organic phosphate lubricant additives. These agents are also used in cutting oils, machine oils, transmission fluids, and cooling lubricants (Lapp, 1976).

In Japan, approximately 300 tonnes of TCP was used for lubricant additives in 1985 (Association of the Plasticizer Industry of Japan, 1985), and approximately 1320 tonnes of TAPs was used in 1976 in Ontario, Canada (Muir et al., 1980).

There are other minor uses of TCP: additives in making synthetic leather (Franchini et al., 1978), shoes (Pegum, 1966), and polyvinyl acetate products (Anon., 1986); solvent for acrylate lacquers and varnishes (Anon., 1986); in non-smudge carbon paper (Hjorth, 1962; Pegum, 1966).

Table 6. Composition of various commercial organophosphorus hydraulic fluids and lubricants

Name	Component (%)			Producer	Reference
	TPP	TCP	Others		
IMOL S-140	1	2 (ortho) 42 (meta) 31 (para)	Tris(dimethylphenyl)- (18) Tris(ethylphenyl)- (6) Tris(trimethylphenyl)- and unknown (1)	Imperial Oil Ltd.	Lockhart et al. (1975)
Pydraul 50E	36		nonylphenyl diphenyl phosphate (40) cumylphenyl diphenyl phosphate (22)	Monsanto Co.	Nevins & Johnson (1978)
Pydraul 115E	7		nonylphenyl diphenyl phosphate (29) cumylphenyl diphenyl phosphate (62)	Monsanto Co.	Deo & Howard (1978)
Pydraul 50E	18.4		nonylphenyl diphenyl (52.8) cumylphenyl diphenyl (24.0)	Monsanto Co.	Deo & Howard (1978)
Kronitex TCP		20.7 (meta) 38.8 (dl-meta, para) 30.4 (dl-para, meta)	dicresyl xylenyl (9.2)	FMC Corp.	Deo & Howard (1978)
Santicizer-140 CDP	14.7	19.4	m-cresyl diphenyl (18.6) p-cresyl diphenyl (14.4) phenyl dicresyl (29.4)	Monsanto Co.	Deo & Howard (1978)
Fyrquel GT	19.2		m-cresyl diphenyl (2.1) phenyl dicresyl (3.2) dicresyl xylenyl (36.2) di(C_3-phenyl) xylenyl (37.1)	Stauffer Chem. Co.	Deo & Howard (1978)
Phosflex 41-P	11.9	40.8	m-cresyl diphenyl (2.1) trixylenyl (9.4) (C_3-phenyl)$_3$ (28.7)	Stauffer Chem. Co.	Deo & Howard (1978)

Table 6 (contd).

Name	Component (%)			Producer	Reference
	TPP	TCP	Others		
Fyrquel 220			phosphates derived from phenol (2.6); o-cresol (0.5); m- and p-cresol (13.6); 2-ethylphenol (0.6) 2,4- and 2,5-xylenol (22.3); mixed xylenol (49.2); 3,4-xylenol (8.6); 6-9 phenolics (1.3); 2,4,6-trimethyl phenol (1.4)	Imperial Oil Ltd.	Pickard et al. (1975)
Kronitex 100	18		diphenyl 2-isopropylphenyl (27) diphenyl 4-isopropylphenyl (11) tris(2-isopropylphenyl) (11) phenyl di-(2-isopropylphenyl) (7) phenyl di-(4-isopropylphenyl) (5)	FMC Corp.	Nobile et al. (1980)
Kronitex 50	33		diphenyl 2-isopropylphenyl (21) diphenyl 4-isopropylphenyl (12) tris(2-isopropylphenyl) (8) phenyl di-(2-isopropylphenyl) (6) phenyl di-(4-isopropylphenyl) (2)	FMC Corp.	Nobile et al. (1980)

4. ENVIRONMENTAL TRANSPORT, DISTRIBUTION, AND TRANSFORMATION

Summary

The majority of TCP release to the environment is accounted for by end-point use rather than production. Total release to the environment in the USA was estimated at 32 800 tonnes in 1977.

TCP released into water is readily adsorbed on to sediment particles, and little or none remains in solution.

TCP is readily biodegraded in sewage sludge with a half-life of 7.5 h, the degradation within 24 h being up to 99%. TCP is almost completely degraded within 5 days in river water. The ortho isomer is degraded slightly faster than the meta or para isomers. Abiotic degradation is slower with a half-life of 96 days.

TCP has, because of its physico-chemical properties, a high potential for bioaccumulation. Laboratory studies of continuous exposure to high concentrations (which are environmentally unrealistic) of radiolabelled TCP have shown high bioconcentration factors (BCF). However, these studies failed to show that the isotope was still associated with the original compound. Taking into account the ready biodegradability of TCP, these data should be viewed as probable overestimates, and it is suggested that little bioaccumulation would occur with environmentally realistic TCP exposure.

4.1 Transport and transformation in the environment

4.1.1 Release to the environment

Total losses of aryl phosphates to the environment in the USA from production in 1977 were estimated as 2585 tonnes, the main source being land disposal of manufacturing wastes (2540 tonnes). Releases from end-product use (32 800 tonnes in the USA in 1977) are estimated to be much greater than from production. The amount of volatilization and leaching from plastic items was 16 300 tonnes and that of leakage of hydraulic fluids and lubricants

13 400 tonnes (Boethling & Cooper, 1985). One major hydraulic fluid manufacturer estimated that as much as 80% of the annual consumption of aryl phosphate hydraulic fluids is used to make up for leakage (MRI, 1979).

No data are available on the release to the atmosphere of TCP from production processes. However, open, high-temperature processes such as roll milling, calendering, and extrusion of plasticized polymers may result in significant gaseous emissions of aryl phosphates (Boethling & Cooper, 1985).

Yasuda (1980) detected significant levels of TCP in urban air and in the atmosphere over coastal waters near industrialized areas. Details are described in section 5.1.1.

The results of a study by the US Environmental Protection Agency (MRI, 1979) showed that TCP can evaporate from automotive upholstery fabric and condense on the interior surface of a relatively cool window.

The emission of TCP from waste incineration plants may also be a pathway to the atmosphere. In a study by Vick et al. (1978), TCP was not detected in vapour samples taken before and after the dust collectors of incinerators.

4.1.2 Fate in water and sediment

The solubility of TCP in water is low (Table 1). Monitoring studies have shown trialkyl/triaryl phosphates to be present in water and sediment sampled near major industrialized areas (Konasewich et al., 1978; Sheldon & Hites, 1978, 1979; Mayer et al., 1981; Williams & Lebel, 1981; Williams et al., 1982; Ishikawa et al., 1985; Fukushima & Kawai, 1986). However, TCP was only occasionally detected in water samples, whereas TPP was often detected (Mayer et al., 1981; Williams & Lebel, 1981; Williams et al., 1982; Ishikawa et al., 1985). The total concentrations of Pydraul (Table 6) components in river (0.24 μg/litre) and lake sediments (570 μg/kg) in the USA revealed a water-sediment difference of more than 3 orders of magnitude (Mayer et al., 1981). Equilibrium of TCP with the bottom sediment in a shallow (0.5-m depth) pond would be expected to be reached rapidly, as in the case of TPP (Muir et al.,

1982). The adsorption coefficient of TCP on marine sediment was found to be 420 (Kenmotsu et al., 1980b).

It is apparent from the following data that TCP is rapidly adsorbed onto river sediment: the level of total aryl phosphate in the sediment of the Kanawha River (USA) was 229 mg/kg at the FMC plant outfall but only 4.4 mg/kg 13 km downstream (Boethling & Cooper, 1985).

Wagemann et al. (1974) and Wagemann (1975) reported that a commercial synthetic lubricating oil, IMOL S-140 (Table 6), degraded in sterilized river water under laboratory fluorescent light and under sunlight at 25 °C, and that the first order rate constant and half-life were, respectively, 9×10^{-3} days^{-1} and 96 days (Wagemann, 1975).

4.1.3 Biodegradation

River die-away studies by Saeger et al. (1979) on nine phosphate esters demonstrated that these esters, exposed to the natural microbial population of the river, underwent primary biodegradation at moderate to rapid rates. A 200-μg portion of TCP was completely degraded within 4 days in 200 ml of Mississippi River (USA) water at room temperature. Hattori et al. (1981) also investigated the degradation of TCP in Neya and Oh River water (Osaka, Japan). After a lag period of 1-2 days, the TCP (1 mg per litre) was almost completely degraded within 5 days under non-sterilized conditions, whereas no degradation in heat-sterilized water occurred during 15 days. In clear non-sterilized sea water, however, the degradation was very slow. Saeger et al. (1979) also found that in sterile river water there was no significant evidence of non-biological degradation or loss. Among the isomers of TCP, the ortho isomer degraded in river water slightly faster than the meta isomer and both isomers degraded faster than the para isomer, which degraded about as fast as TPP (Howard & Deo, 1979).

Primary biodegradation rates from semicontinuous activated sludge (SCAS) studies (US Soap and Detergent Assoc., 1965; Mausner et al.,1969) showed generally the same trend in degradation rates as river die-away studies. At a 24-h feed level of 3-13 mg/litre, TCP showed 99% degradation.

The ultimate biodegradability of TCP was measured by Saeger et al. (1979) using the apparatus and procedure developed by Thompson & Duthie (1968) and modified by Sturm (1973). At 26.4 mg TCP/litre, the carbon dioxide evolution reached 82% of its theoretical value.

For alkyl-aryl and triaryl phosphates, increasing the number and size of substituent groups on the phenyl molecule decreases the biodegradability (Saeger et al., 1979).

The degradation pathway for TCP most probably involves stepwise enzymatic hydrolysis to orthophosphate and phenolic moieties (Barrett et al., 1969; Pickard et al., 1975). The phenol would then be expected to undergo further degradation. Dagley & Patel (1957) demonstrated that p-cresol is oxidized to p-hydroxybenzoic acid by a species of Pseudomonas. Ku & Alvarez (1982) studied the biodegradation of [^{14}C]-tri-p-cresyl phosphate in a laboratory model sewage treatment system, and, in 24-h experiments, found that 70-80% of the TCP (added at 1 mg/litre) was degraded, with a half-life of 7.5 h. The major metabolite extracted with ethyl ether from the aqueous phase was identified as p-hydroxybenzoic acid by thin-layer chromatography and gas chromatography-mass spectrometry, while two other radioactive spots remained unidentified.

4.1.4 Water treatment

Data from FMC Corporation (USA) show that TCP (6.23 mg/litre) in waste water was reduced to 0.23 mg/litre in the effluent water by biological treatment, whereas the aryl phosphates with higher relative molecular mass (>452) (and, therefore, more highly substituted) were not easily removed (Boethling & Cooper, 1985). Fukushima & Kawai (1986) reported that TCP (0.186-9.31 µg/litre) in raw water was reduced to 0.078 µg/litre or less in treated water by conventional waste water treatment. Filtration of effluent samples through 1-µm pore size filters resulted in a further removal of 93% of total aryl phosphates, again demonstrating the adsorptive behavior of these compounds (Boethling & Cooper, 1985).

4.2 Bioaccumulation and biomagnification

4.2.1 Fish

Data on the bioconcentration and depuration of TCP are given in Table 7. None of the exposures were considered to be representative of realistic environmental levels. Moreover the bioconcentration factor (BCF) measured in the laboratory must be considered as a bioaccumulation potential rather than an absolute bioaccumulation factor (Veith et al., 1979).

Several equations have been used in attempts to predict the BCF of organic chemicals in various fish strains using the octanol-water partition coefficient (P_{ow}) or water solubility values (Neely et al., 1974; Lu & Metcalf, 1975; Kanazawa, 1978; Veith et al., 1979; Sasaki et al., 1982).

The clearance of tri-*m*-cresyl phosphate has been shown to be biphasic, with higher rates of clearance in the first 6 days after transfer to clean water, especially for rainbow trout. The clearance rate constants for rainbow trout were about 50% more than those for fathead minnows (Muir et al., 1983).

4.2.2 Plants

The uptake and translocation of tri-*p*-cresyl phosphate by soybean plants has been studied by Casterline et al. (1985), the initial concentration in soil being 10 mg per kg. Approximately 70% of the compound had disappeared from the soil within 90 days (when the plants were harvested). At that time, the amount per plant was 34 µg (0.17% of the applied TCP). Of this total plant content, 74% was found in the stem, 24% in the leaves, and 2% in the pods. The seeds contained no detectable tri-*p*-cresyl phosphate.

Table 7. Bioaccumulation and clearance of tricresyl phosphate by fish

Species	Compound	Flow/stat (temp.)	Analytical method[a]	BCF (K_1/K_2)	Exposure concentration (mg/litre)	Uptake rate (k_1,hr^{-1})	Clearance rate ($k_2 \times 10^{-3}$,h^{-1})	Depuration half-life (hr)	Reference
Rainbow trout (*Salmo gairdneri*)	para isomer	stat(10°C)	TR	2768 ± 641[b]	0.005-0.05		9.6-13.3	72.2	Muir et al. (1983)
			TR	1420 ± 42[c]		14.0			
			TR	1466 ± 138[d]		17.0			
			HER	770 ± 24[c]				65.4	
	meta isomer		TR	1162 ± 313[b]			11.5-24.2	30.3	
			TR	784 ± 82[c]		18.5			
			TR	1102 ± 137[d]		21.2			
			HER	310 ± 52[c]					
Fathead minnows (*Pimephales promelas*)	para isomer	stat(10°C)	TR	2199 ± 227[b]	0.005-0.05		7.0-9.6	25.8	Muir et al. (1983)
			TR	928 ± 78[c]		4.9		90.0	
			TR	588 ± 129[d]		9.6			
			HER	709 ± 76[c]					
	meta isomer		TR	1653 ± 232[b]			8.5-14.7	73.7	
			TR	596 ± 103[c]		7.9		59.2	
			TR	385 ± 92[d]		8.7			
			HER	62 ± 3[c]					
Bluegill (*Leptomis macrochirus*)	commercial TCP	flow(25°C)	GC-FPD	165	0.0316			53.3	Veith et al. (1979)
	para isomer		TR	1589					Sitthichaikasem (1978)

[a] GC-FPD = gas chromatography (flame photometric detector) after suitable extraction; TR = total radioactivity; HER = hexane-extractable radioactivity.
[b] BCF was calculated by the "initial rate method".
[c] The static test method was used (Zitko, 1980).
[d] k_1 and k_2 were derived by non-linear regression calculation.

5. ENVIRONMENTAL LEVELS AND HUMAN EXPOSURE

Summary

TCP has been measured in the atmosphere in Japan at concentrations up to 70 ng/m^3 but only reached 2 ng/m^3 at a production site in the USA. Workplace air in the USA contained less than 0.8 mg/m^3 at a lubrication oil barrel-filling operation and 0.15 mg/m^3 (total phosphates) in an automobile zinc die-casting plant. The concentrations of TCP measured in drinking-water in Canada were low (0.4 to 4.3 ng/litre), and TCP was undetectable in well-water. Levels in river and lake water are frequently considerably higher. However, this is due to the presence of suspended sediment to which TCP is strongly adsorbed. Concentrations up to 1300 µg/kg in river sediment and 2160 µg/kg in marine sediment have been measured.

Elevated TOCP levels in soil and vegetation have been found within the perimeter of production plants.

Residues of TOCP in fish and shellfish up to 40 µg/kg have been reported but the majority of animals sampled contained no detectable amounts.

5.1 Environmental levels

TCP has been found in air, water, soil, sediment, and aquatic organisms. However, the levels of TCP in environmental samples are low (Table 8), except in soil and sediment collected in heavily industrialized areas (Table 9).

5.1.1 Air

Yasuda (1980) measured the distribution of various organic phosphorus compounds in the atmosphere over the eastern Seto Inland Sea, Japan. Near the heavily industrialized cities (Fukuyama, Akashi, Osaka), 11.5-21.4 ng TCP/m^3 was detected. Yasuda (1980) also measured levels of phosphate esters in the atmosphere above the Dogo Plain and the Ozu Basin agricultural area of Western Shikoku. TCP was detected only in the urban air of Matsuyama, where the level was 26.7-70.3 ng/m^3. TCP levels of 0.01-2 ng/m^3 in air collected at production sites in the USA have been reported (MRI, 1979).

Table 8. Concentration of TCP in environmental air, water, sediment, and fish at various locations

Locations	Year	Sample	Concentration	Number of samples (detected/analysed)	Reference
Shikoku (Japan)	1976	atmosphere	26.7-70.3 ng/m^3	(3/19)	Yasuda (1980)
Eastern Seto Inland Sea (Japan)	1977	atmosphere	11.5-21.4 ng/m^3	(3/4)	
Japan (various locations)	1975	river and sea water	ND (50-1500 ng/litre)[a]	(0/100)	EAJ (1977)
		river and sea sediment	150 ng/g	(1/100)	
		fish	ND (20-250 ng/g)[a]	(0/100)	
Japan (various locations)	1978	river and sea water	ND (50-2500 ng/litre)[a]	(0/114)	EAJ (1979)
		river and sea sediment	1060-2160 ng/g	(3/114)	
		fish	ND (0.25-150 ng/g)[a]	(0/98)	
Osaka (Japan)	1976	river water	100-9500 ng/litre	(11/13)	Kawai et al. (1978)
Eastern Ontario water treatment plant (Canada)	1978	drinking-water	0.3 ng/litre	(1/12)	Lebel et al. (1981)
Tokyo (Japan)	1978	river water	ND (50 µg/litre)[a]	(0/12)	Wakabayashi (1980)
		sea water	ND (50 µg/litre)[a]	(0/3)	
		river sediment	7-370 ng/g	(9/10)	
		sea sediment	4 ng/g	(1/3)	
Canada (various locations)	1979	drinking-water	0.7-4.3 ng/litre	(7/60)	Williams & Lebel (1981)
Great Lake (Canada)	1980	drinking-water	0.4-1.8 ng/litre	(5/12)	Williams et al. (1982)
Kitakyushu City (Japan)	1980	river water	67-259 ng/litre	(3/16)	Ishikawa et al. (1985)
		sea water	ND (20 ng/litre)[a]	(0/9)	
		sea sediment	ND (10 ng/g)[a]	(0/6)	
Seto Inland Sea (Japan)	1980	fish and shell fish	1-19 ng/g	(4/41)	Kenmotsu et al. (1981a)

[a] Range of detection limits due to analytical methods used; ND = not detected.

Table 9. Concentration of TCP detected near the producers and users of trialkyl/aryl phosphates

Locations	Year	Sample	Concentration	Number of samples (detected/analysed)	Reference
Near TAPs manufacturing plants (USA)		fish	2-5 ng/g		Muir (1984)
Columbia River (USA)		fish (sturgeon)	40 ng/g		Lombardo & Egry (1979)
Kanawha River (USA)	1978	river water	20 000 ng/litre		Boethling & Cooper (1985)
FMC Corp., Nitro, MV (USA)	1979	air (HV)[a]	2 ng/m^3		Boethling & Cooper (1985)
Stauffer Chemical Co., Gallipolis Ferry, MV (USA)	1979	air (HV)[a] vegetation soil	0.01-0.05 ng/m^3 1000-20 000 ng/g 1000-4000 ng/g	(2/4) (4/4) (4/4)	Boethling & Cooper (1985)
FMC Corp. Plant (USA)	1980	waste water effluent water	6.23 mg/litre 0.23 mg/litre		Boethling & Cooper (1985)
Baltimore Harbour (USA)	1983	sediment	400-600 ng/g	(2/3)	Boethling & Cooper (1985)
Detroit River, mouth (USA)	1983	sediment	230-1300 ng/g	(2/2)	Boethling & Cooper (1985)

[a] HV = High volume filter pad (air sampler).

Environmental Levels and Human Exposure

5.1.2 Water

Although there have been many monitoring studies for TAPs in water, TCP has not often been detected in natural water. Where present, it is only at low levels. According to the annual reports of the Environment Agency of Japan, TCP has not been detected in river or sea water at any sampling points. Due to the variety of analytical methods and procedures used, the detection limits varied between 5 and 2500 ng/litre between different laboratories (EAJ, 1977; 1979; 1981). Kawai et al. (1978) detected TCP at 100-9500 ng/litre in river water sampled in Osaka (Japan), and found that the concentration of TCP in river water tended to parallel the concentration of suspended solid. Ishikawa et al. (1985) detected TCP levels of 67-259 ng/litre in 3 out of 16 samples of river water in Kitakyushu City (Japan), but not in sea water. Both cities are located in the most heavily industrialized areas of Japan.

Relatively high concentrations of TAPs have frequently been detected in river water sampled near producer or user sites: 20 μg TCP/litre was detected in the Kanawha River (USA) 8 miles downstream from the plant outfall (Boethling & Cooper, 1985).

5.1.3 Soil

There has been only one report of TCP in soil (from Stauffer Chemical Co. at Gallipolis Ferry (USA)), the level being 1.0-4.0 mg/kg (Boethling & Cooper, 1985). The high concentration of total TAPs (26 550 mg/kg) in this sample was thought to reflect product accumulation in the area, which was subject to frequent spills.

5.1.4 Sediment

Because of the high sediment adsorption coefficient, higher levels of TCP have frequently been detected in sediment than in water. TCP was detected at 400-600 ng/g in sediment in Baltimore harbour (USA) and at 230-1300 ng/g in the Detroit River (USA) (Boethling & Cooper, 1985). According to the annual reports of the Environment Agency of Japan, a level of 150 ng/g was found (Mitaki

River, Japan) in one out of 100 sediment samples in 1975, whereas 1060-2160 ng/g (Doukai Bay, Japan) was found in 3 out of 114 samples in 1978 (EAJ, 1977; 1979; 1981). Wakabayashi (1980) detected 7-370 ng/g in nine out of ten river sediment samples, and 4 ng/g in one out of three sea sediment samples in Tokyo.

5.2 General population exposure

5.2.1 Drinking-water

TCP levels in drinking-water are very low. Lebel et al. (1981) analysed TAPs in drinking-water sampled from eastern Ontario water treatment plants and found TCP (at 0.3 ng/litre) in only one out of 12 samples. In an extended survey of drinking-water conducted in Canada (Williams & Lebel, 1981), TCP was detected at 0.7-4.3 ng/litre in 7 out of 60 samples of treated potable water obtained at the treatment plants of 29 municipalities.

In a study by Williams et al. (1982), TCP was detected in river and lake water but not in well-water. TCP was also found, at concentrations of 0.4 to 1.8 ng/litre, in 5 out of 12 samples of drinking-water obtained from 12 water treatment plants located around the Great Lakes (USA).

In general, the TCP concentration in drinking-water is lower (by factors of 10^{-2} to 10^{-3}) than that in river water. This is due to the efficient removal of TCP at water treatment plants by infiltration using activated carbon with a high adsorption coefficient.

5.2.2 Fish

Lombardo & Egry (1979) found a TCP concentration of 40 ng/g in sturgeon caught in the Columbia River (USA), where many metal-processing plants were located upstream from the sampling point. Muir (1984) found 2-5 ng/g in fish caught near TAP manufacturing plants. According to the annual reports of the Environment Agency of Japan, TCP was not detected in fish caught at any sampling points (EAJ, 1977; 1979; 1981). The analytical detection limits varied from 0.25 to 250 ng/g. Kenmotsu et al. (1981a) found 1-19 ng/g in 4 out of 41 samples of fish and shellfish collected in the Seto Inland Sea, Japan.

5.2.3 Human tissues

There has been only one report of TAPs in human adipose tissues (Lebel & Williams, 1983). Although there was no history of TCP exposure in these patients, tris(1,3-dichloroisopropyl) phosphate and tributoxyethyl phosphate were detected at levels of 0.5-110 ng/g and 4.0-26.8 ng/g, respectively.

5.3 Occupational exposure

The National Institute for Occupational Safety and Health, USA (US NIOSH) has monitored workplace air, and found that air samples collected near barrel-filling operations where lubricating oil was produced by blending TCP contained less than 0.8 mg TAP/m^3 (US NIOSH, 1979). Air collected near the zinc die-casting machine in automobile manufacturing contained a total phosphate ester level of 0.15 mg/m^3 (US NIOSH, 1980). Airborne TOCP resulting from the production of heavy-duty radiators has been investigated by NIOSH, but the concentration was below the limit of detection (US NIOSH, 1982). Triaryl phosphates (at approximately 0.1 ppm) were detected in the air of the aircraft elevator machinery spaces on the carrier USS Leyte (CVS-32) where a triaryl phosphate oil was used as a hydraulic fluid (Baldridge et al., 1959).

6. EFFECTS ON ORGANISMS IN THE ENVIRONMENT

Summary

The primary productivity of cultures of freshwater green algae was reduced to 50% by TOCP at levels of 1.5 to 4.2 mg per litre, depending on the species. The meta and para isomers were less toxic. There are few data on the acute toxicity of TCP to aquatic invertebrates: a 48-h LC_{50} for Daphnia of 5.6 mg per litre, a 24-h LC_{50} for a nematode of 400 mg/litre, and a 2-week NOEL for Daphnia (mortality, growth, reproduction) of 0.1 mg/litre. The 96-h LC_{50} values for three fish species varied between 4.0 and 8700 mg TCP/litre. Rainbow trout showed approximately 30% mortality after a 4-month exposure to IMOL S-140 (2% TOCP) at 0.9 mg/litre and minor effects within 14 days. The exposure levels used in these studies were much higher than likely water concentrations in the environment and, in most cases, greatly exceeded the solubility of the compounds.

There is no information on the bioavailability or toxicity to burrowing or bottom-living organisms of TCP bound to sediment.

There is an indication that crop plants can be affected by TOCP released from plastic coverings, but there is no information concerning the effects on wild plant species.

6.1 Unicellular algae

Data on the toxicity of TCP compounds to unicellular algae are given in Table 10.

The toxicity of TCP compounds to freshwater algae depends on their chemical structure. Substitution of the hydrogen by a methyl group in the benzene ring decreases the toxicity (Wong & Chau, 1984). Of the TCP isomers, the ortho isomer was the most toxic for the primary productivity of *Ankistradesmus falcatus*, followed by the meta and para isomers (Wong & Chau, 1984).

6.2 Aquatic organisms

Data on the toxicity of TCP to aquatic organisms are presented in Table 11.

Table 10. Toxicity of TCP to freshwater unicellular algae

Organism	Isomer	Temperature (°C)	Species	Effect	Concentration (mg/litre)	Reference
Alga	ortho	20	Ankistrodesmus falcatus var. acicularis	24-h IC_{50} for primary productivity	2.5	Wong & Chau (1984)
	meta	20			> 5.0	
	para	20			> 5.0	
Green alga	ortho	20	Scenedesmus quadricaudata	24-h IC_{50} for primary productivity	4.2	Wong & Chau (1984)
	meta	20			> 5.0	
	para	20			> 5.0	
Lake Ontario phytoplankton	ortho	20		24-h IC_{50} for primary productivity	1.7	Wong & Chau (1984)
	meta	20			4.1	
	para	20			> 5.0	
Green alga			Scenedesmus pannonicus	4-day EC_{50}	1.5	Adema et al. (1983)[a]

[a] Tests were performed according to or in line with standardized procedures (OECD, 1981). EC_{50}: 50% effective concentration; IC_{50}: 50% inhibition concentration.

Table 11. Toxicity of TCP to aquatic organisms

Organisms	Chemicals	Size/weight	Temp. (°C)	Flow/stat	Hardness (mg/litre)	End-point or criteria	Parameter	Concentration (mg/litre)	Reference
Tidewater silverside (*Menidia beryllina*)	TCP[a]	40-100 mm	20	stat			96-h LC$_{50}$	8700	Dawson et al. (1977)
Bluegill (*Lepomis macrochirus*)	TCP[a]	35-75 mm	23		55		96-h LC$_{50}$	7000	Dawson et al. (1977)
		0.60 g	12	flow-through	44 314		96-h LC$_{50}$	0.26 0.061	Mayer & Ellersieck (1986)
Guppy (*Poecilia reticulata*)	TCP[a]			stat		mortality, swimming behaviour, colour	96-h LC$_{50}$ 96-h NOEL	4.0 1.0	Adema et al. (1983)[b] Adema et al. (1983)[b]
						mortality, growth, swimming behaviour	4-week NOEL	1.0	Adema et al. (1983)[b]
	IMOL S-140			stat		mortality, visible effects	24-h NOEL	> 57	Wagemann (1975)
Flagfish (*Jordanella floridae*)	TCP[a]					egg-larval development, mortality, growth, swimming behaviour, colour	6-week NOEL	0.01	Adema et al. (1983)[b]

Table 11 (contd).

Organisms	Chemicals	Size/weight	Temp. (°C)	Flow/stat	Hardness (mg/litre)	End-point or criteria	Parameter	Concentration (mg/litre)	Reference
Rainbow trout (*Salmo gairdneri*)	IMOL S-140			flow-through		ate less, less active	condition after 8 days	0.9	Wagemann (1975)
						ceased surface feeding	condition after 14 days	0.9	Wagemann (1975)
						mortality (5/16)	condition after 4 months	0.9	Wagemann (1975)
	TCP[a]	0.23 g 0.50 g	12	flow-through	44		96-h LC$_{50}$	0.26 0.40	Mayer & Ellersieck (1986)
Waterflea (*Daphnia magna*)	TCP[a]			flow-through		mortality	48-h LC$_{50}$	5.6	Adema et al. (1983)[b]
						mortality, reproduction, growth	2-week NOEL	0.1	Adema et al. (1983)[b]
Channel catfish (*Ictalurus punctatus*)		1.30 g	12	flow-through	44		96-h LC$_{50}$	0.80	Mayer & Ellersieck (1986)
Yellow perch		0.79 g			292				

[a] No descriptions of isomeric compositions were given in the references.
[b] Tests were performed according to or in line with standardized procedures (OECD, 1981).

Measurement of the acute toxicity (96-h LC_{50}) of TCP to fish range from 8700 mg/litre in tidewater silversides (Dawson et al., 1977) to 4 mg/litre in guppies (Adema et al., 1983). The composition of the materials that were used in these experiments was not given.

Tests on guppies showed that a saturated solution of IMOL S-140 (see Table 6) in water (14 mg/litre) was not acutely toxic (96-h exposure), but exposures of 4 months or more at concentrations of 0.3-0.9 mg/litre caused symptoms of chronic poisoning in rainbow trout. Initially only feeding habits and behaviour changed, but later swimming ability was impaired, and the fish eventually died (Wagemann et al., 1974; Wagemann, 1975). Fatty tissue turned a blue-grey colour, the liver enlargened, and the activities of lactate dehydrogenase (LDH) and glutamic-oxaloacetic transaminase (GOT) increased (Wagemann et al., 1974; Wagemann, 1975; Lockhart et al., 1975).

Phosflex 179-C (TOCP) slightly inhibited the acetyl-cholinesterase activity of an electric ray *(Torpedo electroplax)* but did not interfere with binding of acetylcholine to its receptor (Eldefrawi et al., 1977).

Fish or frogs that received IMOL S-140 or TOCP did not show, under the test conditions, significant reduction of brain cholinesterase activity (Lockhart et al., 1975). A similar observation was made by Cohen & Murphy (1970) on mice and quail.

6.3 Insects

The toxicity of TCP to insects is presented in Table 12. Most of these data were obtained in the course of studies on the synergism of TCP or TPP with organophosphorus insecticides or juvenile insect hormone mimics.

6.4 Plants

The effects of gaseous TCP on crops covered with vinyl film have been investigated. TCP emitted from the film caused a certain amount of leaf shrinking (Inden & Tachibana, 1975).

The active metabolite of TOCP, saligenin cyclic *o*-tolyl phosphate, caused decreased germination of kidney beans and wheat (Eto et al., 1962).

Table 12. Toxicity of TCP to insects

Species	TCP isomer	Application method	Age	Effect[a]	Concentration	Reference
Mosquito larva (Aedes aegypti)	ortho	in water	early 4th instar	5-day LD_{13} [± 19] (LD_2 [± 3] in control)	0.1 mg/litre	Quinstad et al. (1975)
Mosquito larva (Culex pipiens quinquefasciatus)		in water	4th instar	5-day LD_7 [± 7] (LD_5 [± 5] in control)	0.1 mg/litre	Quinstad et al. (1975)
Mosquito larva (Culex tarsalis)		in water	4-h instar	24-h LC_{50}	> 1 mg/litre	Plapp & Tong (1966)
Housefly larva (Musca domestica)		topical treatment	3rd instar	7-day LD_{12} [± 6] (LD_8 [± 10] in control)	0.1 mg/g	Quinstad et al. (1975)
Housefly (Musca domestica)		contact method	2–5 days old	24-h LD_{50}	> 1 mg/jar	Plapp & Tong (1966)
Housefly (Musca domestica)	meta	contact method	2–5 days old	24-h LD_{50}	> 1 mg/jar	Plapp & Tong (1966)
Mosquito larva (Culex tarsalis)		in water	4th instar	24-h LD_{50}	> 1 mg/litre	Plapp & Tong (1966)
Housefly (Musca domestica)	para	contact method	2–5 days old	24-h LD_{50}	> 1 mg/jar	Plapp & Tong (1966)
Mosquito larva (Culex tarsalis)		in water	4th instar	24-h LD_{50}	> 1 mg/litre	Plapp & Tong (1966)

[a] Values in square brackets are standard deviations.

7. KINETICS AND METABOLISM

Summary

The absorption, distribution, metabolism, and elimination of organophosphates are critical to the delayed neuropathic effects of these compounds. In addition, other factors (e.g., route of administration, sex, age, strain) affect their metabolic fate and subsequent neurotoxic expression. Variability in these factors may underline the interspecies variation in the sensitivity to TOCP-induced delayed neuropathy (i.e. OPIDN). This correlation has been demonstrated with other OPIDN compounds, but relevant studies on TOCP itself are limited.

Dermal absorption of TOCP in humans appears to be at least an order of magnitude faster than in dogs. Significant dermal absorption also appears to occur in cats. Oral absorption of the compound has been reported in rabbits. There is no direct information on absorption via the inhalation route.

In cat studies, absorbed TOCP was widely distributed throughout the body, the highest concentration being found in the sciatic nerve, a target tissue. Other tissues with high concentrations of TOCP and its metabolites are the liver, kidney, and gall bladder.

TOCP is metabolized via three pathways. The first is the hydroxylation of one or more of the methyl groups, and the second is dearylation of the o-cresyl groups. The third is further oxidation of the hydroxymethyl to aldehyde and carboxylic acid. The hydroxylation step is critical because the hydroxymethyl TOCP is cyclized to form saligenin cyclic o-tolyl phosphate, the relatively unstable neurotoxic metabolite.

TOCP and its metabolites are eliminated via the urine and faeces, together with small amounts in the expired air.

7.1 Absorption

TOCP absorption has been studied in a variety of species using oral or dermal administration. No information is available on absorption following inhalation.

Gross & Grosse (1932) reported that TOCP given orally (0.1 g/kg in olive oil) was absorbed by rabbits.

Hodge & Sterner (1943) demonstrated poor absorption of ^{32}P-labelled TOCP in a dog after administration of a single dermal dose of 200 mg/kg. The rate of transfer (dose: 2-4 mg [^{32}P]-TOCP/kg) through intact human palm skin appeared to be about 100 times faster than that through the abdominal skin of the dog. This was based on urinary excretion and surface area considerations.

Another species, the cat, showed even greater absorption. When [^{14}C]-TOCP (50 mg/kg) was dermally applied to adult male cats, the disappearance of radioactivity from the application site was bi-exponential. In the first phase, 73% of the TOCP disappeared within 12 h, while the second phase half-life was 2 days (Nomeir & Abou-Donia, 1984; 1986b).

Studies by Kurebayashi et al. (1985) indicated incomplete absorption of tri-p-cresyl phosphate (TPCP) from the intestine of rats after a single oral dose of [methyl-^{14}C]-TPCP (7.8 or 89.6 mg/kg) in 1.5 ml of dimethyl sulfoxide. Much of the radioactivity was recovered in the faeces, predominantly in the form of unchanged TCP.

7.2 Distribution

After a single oral dose of [^{32}P]-TOCP (770 mg/kg) to chickens, the total radioactivity in liver increased consistently throughout 72 h. The levels of radioactivity in the plasma were consistently lower than those in liver; at 24 h the plasma levels were 5% of those in liver. The radioactivity was predominantly associated with TOCP metabolites in liver but with unmetabolized TOCP in blood (Sharma & Watanabe, 1974).

Following a single dose of [^{32}P]-TOCP (200 mg/kg) to the abdominal skin of the dog, the radioactivity in the blood within 24 h was equivalent to an average value of 80 µg/litre and was distributed throughout the visceral organs, muscle, brain, and bone. The levels of radioactivity in tissues were in the following descending order:

liver > blood > kidney > lung > muscle or spinal cord > brain or sciatic nerve (Hodge & Sterner, 1943).

In cats given a single dermal dose of 50 mg [^{14}C]-TOCP/kg, the chemical was absorbed from the skin and subsequently distributed throughout the body. TOCP reached its highest concentration in plasma at 12 h, and its metabolites attained their maximum concentration between 24-48 h. The relative residence values of unmetabolized TOCP in various tissues, relative to the plasma, were: brain, 0.09; spinal cord, 0.18; sciatic nerve, 2.1; liver, 0.44; kidney, 0.55; lung, 1.27. Parent TOCP was the predominant compound in the brain, spinal cord, and sciatic nerve, while the metabolites o-hydroxybenzoic acid and di-o-cresyl phosphate were predominant in the liver, kidney, and lung (Nomeir & Abou-Donia, 1984). In contrast, when measuring total radioactivity sampled 1-10 days post exposure, highest levels were found in the bile, gall bladder, urinary bladder, kidney, and liver, with only low levels in the spinal cord and brain (Nomeir & Abou-Donia, 1986b).

Gross & Grosse (1932) reported that most of the cresol ester was recovered from the liver (5%) and intestine (67%) within 2 h after an intravenous injection (0.5 g/kg) of TOCP into rabbits.

At 24, 72, and 168 h after oral administration of [^{14}C]-TPCP to rats, the concentrations of radioactivity in adipose tissue, liver, and kidney were higher than those in other tissues (Kurebayashi et al., 1985).

7.3 Metabolic transformation

TOCP is metabolized in rats, rabbits, mice, and chickens to form a neurotoxic esterase inhibitor (Davison, 1953; Aldridge, 1954; Aldridge & Barnes, 1961; Casida, 1961). In rats injected intraperitoneally with TOCP, this esterase inhibitor was located mainly in the intestine and liver (Myers et al., 1955). The neurotoxic metabolite was isolated from the intestine and liver of rats following TOCP administration and was identified as saligenin cyclic o-tolyl phosphate [2-(o-cresyl)-4H-1:3:2-benzodioxaphosphoran-2-one] (M-1); M2 and M3 in Fig. 1 are also possible metabolites (Casida et al., 1961; Eto et al., 1962). The saligenin cyclic o-tolyl phosphate was also found in chickens (Eto et al., 1962; Sharma & Watanabe, 1974) and in cats (Taylor & Buttar, 1967; Nomeir & Abou-Donia, 1984; 1986a,b). Although quantitative data are not available, indirect evidence suggests that cats metabolize this

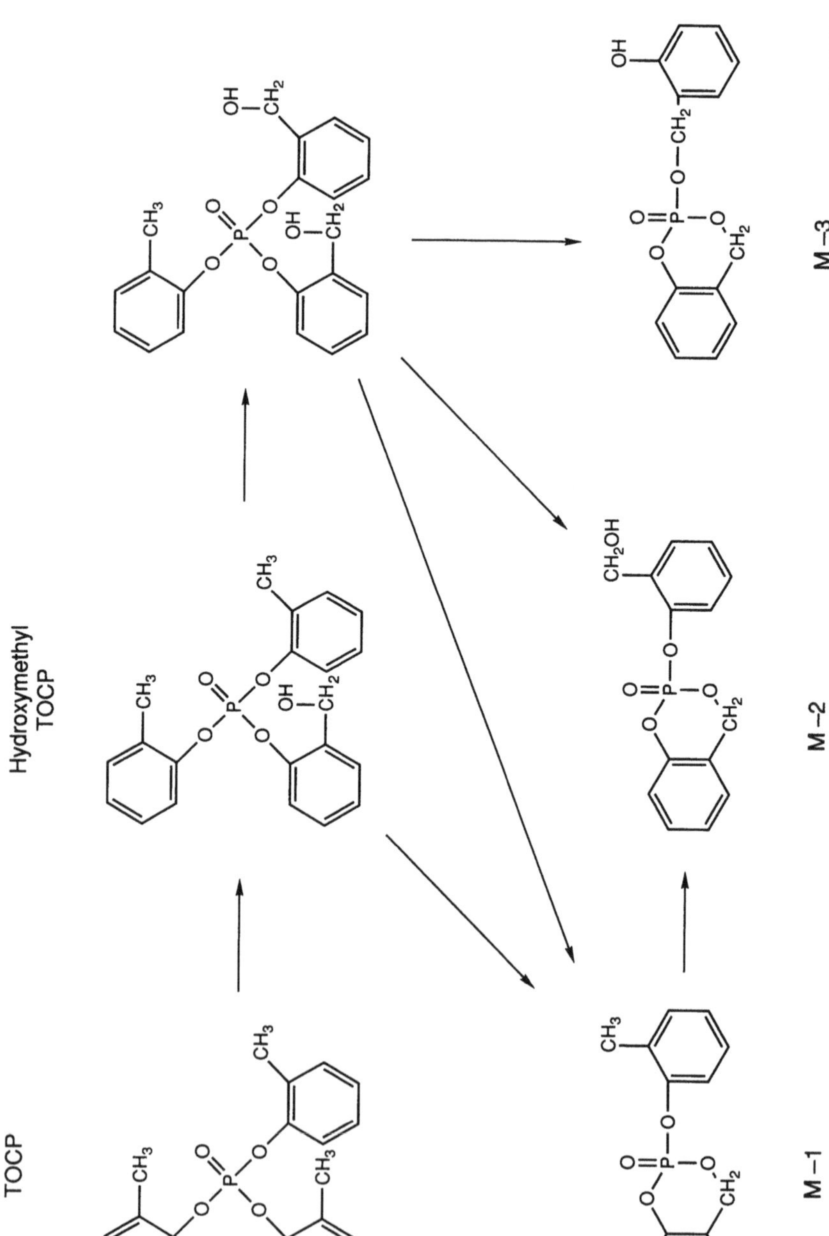

Fig. 1. Proposed pathway for hydroxylation and cyclization reactions in the metabolism of TOCP.

neurotoxic compound more efficiently than chickens (Taylor & Buttar, 1967). Two intermediate metabolites, di-(o-cresyl) mono-o-hydroxymethylphenyl phosphate [monohydroxymethyl TOCP] and di-(o-hydroxymethylphenyl) mono-o-cresyl phosphate, [di-hydroxymethyl TOCP], transform to saligenin cyclic o-tolyl phosphate (Eto et at., 1962; 1967), which is relatively unstable and is rapidly hydrolysed to inactive metabolic products.

TOCP is metabolized via three essential pathways. The first is the hydroxylation of one or more of the methyl groups to hydroxymethyl, which is responsible for the formation of mono- and di-hydroxymethyl TOCP and o-hydroxybenzyl alcohol. This reaction is known to be catalysed by the microsomal mixed-function oxidase system (Eto et al., 1967). The hydroxymethyl TOCP is cyclized to form saligenin cyclic o-tolyl phosphate with spontaneous release of o-cresol, this being catalysed by the reaction of plasma albumin or other components (Eto et al., 1967). The cyclic phosphate metabolite is relatively unstable and is rapidly hydrolysed to inactive metabolic products (Eto et al., 1967). The second pathway is the dearylation of one or more of the o-cresyl groups of TOCP, resulting in the formation of o-cresol, di-o-cresyl phosphate, o-cresyl phosphate, and phosphoric acid. The third pathway is further oxidation of hydroxymethyl to aldehyde and carboxylic acid. These oxidation reactions are most likely to be mediated by alcohol and aldehyde dehydrogenases.

Studies with [^{32}P]-TOCP in rats have shown that hydrolysis leads to the rapid excretion in the urine of diaryl phosphates, monoaryl phosphates, and phosphoric acid (Casida et al., 1961).

Nomeir & Abou-Donia (1984; 1986a,b) clearly identified the metabolites of TOCP in male cats. Mono- and di-hydroxymethyl TOCPs and saligenin cyclic-o-tolyl phosphates were present in most tissues, but their concentrations were low compared with those of other metabolites. The major metabolite of TOCP in the liver, kidney, lung, and urine of cats was o-hydroxybenzoic acid; di-o-cresyl phosphate, o-cresyl phosphate, o-cresol, o-hydroxybenzyl alcohol, and o-hydroxybenzaldehyde were also identified. However, the brain, spinal cord, sciatic nerve, and faeces contained predominantly unchanged TOCP.

Johnson (1975a) compared the metabolic pathways of the three isomers of TCP. The main observations, which concerned several organophosphorus esters, were as follows:

(i) Provided that the *o*-alkyl group has at least one hydrogen on the α-carbon atom, cyclic derivatives can be obtained that are often highly neurotoxic.

(ii) At the para position, a substituent requires two hydrogen atoms on the α-carbon atom in order to produce a neurotoxic metabolite inhibiting NTE.

(iii) Substituents at the meta position may be metabolized but do not yield inhibitory products.

The major urinary metabolites of TPCP in rats were *p*-hydroxybenzoic acid, di-*p*-cresyl phosphate, and *p*-cresyl *p*-carboxyphenyl phosphate. Mono-(or di-)*p*-cresyl di-(or mono-)*p*-carboxyphenyl phosphate was identified as the intermediate metabolite in the bile (Kurebayashi et al., 1985).

7.4 Excretion

After a single oral dose of [^{32}P]-TOCP (770 mg/kg) to hens, 26.5% of the total radioactivity was eliminated in the combined urinary-faecal excreta over 72 h, mostly as TOCP (Sharma & Watanabe, 1974).

After a single dose of [^{14}C]-TOCP (50 mg/kg) to male cats, approximately 28% of the applied dose was excreted in the urine and 20% via the bile into the faeces within 10 days (Nomeir & Abou-Donia, 1986b). After this exposure, the disappearance of TOCP and its metabolites from the plasma followed monoexponential kinetics. The apparent half-lives of TOCP and its metabolites (in days) in the plasma were: TOCP, 1.20; saligenin cyclic-*o*-tolyl phosphate, 2.47; di-*o*-cresyl phosphate, 4.50; *o*-cresyl phosphate, 4.30; *o*-cresol, 2.65; *o*-hydroxybenzyl alcohol, 14.0; *o*-hydroxybenzaldehyde, 5.70; *o*-hydroxybenzoic acid, 6.00; monohydroxymethyl TOCP, 2.20. The apparent half-lives of TOCP and its metabolites reflected the rates of all processes involving the conversion, clearance, and/or redistribution of these metabolites (Nomeir & Abou-Donia, 1984).

Elimination via the bile has been demonstrated after intravenous injection into rabbits (Gross & Grosse, 1932) and intraperitoneal injection into rats (Myers et al., 1955). Smith et al. (1932) measured the urinary phenol excretion in cats given subcutaneous doses of 0.4 to 1.0 ml TOCP/kg. Little if any increase in the urinary phenol excretion was found either before or after the onset of paralysis of the hindlimbs.

After a single oral dose (500 mg/kg) of tri-*m*-cresyl phosphate (TMCP) or TPCP to rabbits, 92% of TMCP and 95% of TPCP was eliminated in the faeces within 4 days (Gross & Grosse, 1932). After a single oral dose of [methyl-^{14}C]-TPCP (7.8 or 89.6 mg/kg), about 90% or 76%, respectively, of the radioactivity was eliminated in the urine and faeces within 24 h (Kurebayashi et al., 1985). The apparent half-lives of the radioactivity in tissues ranged from 14 h for blood to 26 h for lung and brain. At the lower dose level, about 28% of the dose was eliminated via the bile within 24 h. The expiratory excretion as $^{14}CO_2$ over 3 days amounted to 18% of the radioactivity, which was reduced to 3% when the rats were treated with neomycin. The authors suggested that the enterohepatic circulation and intestinal microflora play an important role in the degradation of TPCP biliary metabolites.

8. EFFECTS ON EXPERIMENTAL ANIMALS AND *IN VITRO* TEST SYSTEMS

Summary

Of the three isomers of TCP, TOCP is by far the most toxic and is the only isomer that produces delayed neuropathy.

A wide animal interspecies variability exists for the various end-points (e.g., lethality, neuropathology, ataxia, enzyme inhibition) of TOCP exposure, the chicken being one of the most sensitive species to delayed neuropathy (i.e. OPIDN).

Species sensitivity to the lethal effects of TOCP administration is highly variable, chickens and cats being more sensitive than rats and mice. A single oral dose of 50-500 mg TOCP per kg induced delayed neuropathy in chickens, whereas doses of 840 mg/kg or more were necessary to produce spinal cord degeneration in Long-Evans rats.

No effects were reported in skin and eye irritation studies on rabbits.

Reproduction studies on rats and mice receiving repeated oral exposure to TOCP showed histopathological damage in the testes and ovaries, morphological changes in sperm, decreased fertility in both sexes, and decreased litter size and viability. However, no reproductive effects were seen in a study with TPCP. A clear no-observed-effect level for the reproductive effects of TOCP was not apparent from the data available. A study in rats, using oral doses that produced maternal toxicity, failed to show any teratogenicity.

Little or no information is available on mutagenicity and carcinogenicity.

Delayed neuropathy has been produced with both single and repeated exposure regimes in a wide range of experimental species, and it is classified as a "dying-back neuropathy". "Neurotoxic esterase" is thought to be the biochemical target of OPIDN and its inhibition by more than 65% shortly after exposure to TOCP presages subsequent neuropathy. Factors other than metabolism (e.g., route of exposure, age, sex, species, strain) influence variability in sensitivity to OPIDN.

Electrophysiology studies have been performed in cats and chickens exposed to TOCP.

In the chicken, single exposures below 58 mg/kg or short-term (i.e. 90-day exposure) daily doses of less than 5 mg/kg appear to be no-observed-effect levels for delayed neuropathy.

8.1 Single exposure

The acute toxicity of TCP to different species is summarized in Table 13.

Table 13. LD_{50} values for TCP and its isomers

Compounds	Route of admin.	Species	LD_{50} (mg/kg)	Reference
Tricresyl phosphate (mixed isomers)	oral	rat	5190	Marhold (1972)
	oral	rat	> 4640	Stauffer (1988)[a]
	oral	rat	> 15 800	Johannsen (1977)
	oral	mouse	3900	Izmerov (1982)
	oral	chicken	> 10 000	Johannsen et al. (1977)
	dermal	rabbit	> 7900	Johannsen et al. (1977)
	dermal	cat	1500	Abou-Donia et al. (1980)
Tri-o-cresyl phosphate	oral	rat	8400	Johannsen et al. (1977)
	oral	rat	1160	Veronesi et al. (1984a)
	oral	rabbit	3700	Johannsen et al. (1977)
	oral	chicken	500	Kimmerle & Loeser (1974)
	oral	chicken	100-200	Smith et al. (1932)
Tri-p-cresyl phosphate	oral	rabbit	> 3000	Smith et al. (1932)
	oral	chicken	> 1000	Smith et al. (1932)
Tri-m-cresyl phosphate	oral	rabbit	> 3000	Smith et al. (1932)
	oral	chicken	> 2000	Smith et al. (1932)

[a] Personal communication to the IPCS from Stauffer Chem. Co. (1988) entitled: Test procedures and data summaries for t-butyl phenyl diphenyl phosphate, tricresyl phosphate, trixylenyl phosphate, mixed triaryl phosphate and isopropyl phenyl diphenyl.

The acute symptoms of intoxication are typical of organophosphorus poisoning. The most toxic compound appears to be TOCP; the acute toxicity of TCP depends on the relative proportions of the different isomers.

The chicken, guinea-pig, and rabbit are the most sensitive species, death occurring at an oral dose of 100 mg/kg (Smith et al., 1932); rats and mice are the least sensitive species.

Sheep given oral doses (100, 200, or 400 mg/kg) of TOCP exhibited acute intoxication characterized by diarrhoea, dehydration, metabolic acidosis, and death within 6 days. Pigs dosed with 100 to 1600 mg TOCP/kg showed minimal signs of acute intoxication, but developed severe signs of delayed neuropathy approximately 15 days after administration (Wilson et al., 1982).

8.2 Short-term exposure

Saito et al. (1974) conducted a 3-month study in rats with TCP consisting of 60-65% TMCP and 35-40% TPCP. The compound was suspended in water with 5% gum arabic and given orally to SD rats at 30, 100, 300, or 1000 mg/kg per day. Histopathological examination revealed no notable changes associated with the compound. Based on those observations, the authors concluded that TCP was of low short-term toxicity.

Oishi et al. (1982) reported that Wistar rats fed a pellet diet containing a mixture of TCP isomers at a concentration of 5 g/kg diet for 9 weeks developed increases in absolute and relative liver weights. Haematological examination revealed no notable changes, but, in the plasma, total protein, urea, cholesterol and glutamate-pyruvate transaminase were significantly increased. Slight liver histopathology included cytoplasmic vacuolization, increase in the number of binucleated cells, and enlargement of cell size.

Chapin et al. (1988) exposed male and female CD-1 mice to diets containing 0, 0.437, 0.875, 1.75, 3.5, or 7.0% TCP (a mixture of ortho, meta, and para isomers) for 14 days. No clinical signs of toxicity were observed in the animals at doses up to 0.875%. All animals in the groups given 1.75, 3.5, or 7.0% exhibited piloerection, tremors, and diarrhoea, and were lethargic before death during the 14-day exposure.

8.3 Skin and eye irritation

No published information is available.

8.4 Teratogenicity

Mele & Jensh (1977) reported that no abnormalities were found in fetuses from pregnant Wistar rats treated with 500 or 750 mg TOCP/kg of on the 18th and 19th days of gestation. This study was primarily designed to investigate the effects of prenatal treatment with TOCP on postnatal behaviour. As such, it cannot be regarded as a true teratogenicity study.

Tocco et al. (1987) tested the teratogenicity of TOCP in Long-Evans rats treated with 87.5, 175, and 350 mg/kg per day throughout organogenesis from gestation days 6 to 18. No maternal deaths or toxicity were observed at the low or medium dose levels. Maternal lethality in the high-dose group was higher than that in the control groups. Numerous soft tissue and skeletal malformations were observed in both control and TOCP-treated groups, but there were no significant differences in the frequency of malformations between the treated and control animals.

8.5 Reproduction

In studies by Somkuti et al. (1987a), TOCP was tested for effects on the male reproductive tract in male Fisher-344 rats. Animals were dosed daily for 63 days at dose levels ranging from 10 to 100 mg/kg per day. Vehicle-treated animals served as controls. As judged by enzymatic, hormonal, and sperm motility, density, and morphology investigations, the minimum effective (threshold) dose for observable testicular toxicity was 10-25 mg/kg per day. The data suggested that TOCP interfered directly with spermatogenic processes and sperm motility and not via androgenic mechanisms or decreased vitamin E availability. Testicular pathological changes were seen at doses above 25 mg/kg per day and included the following: PAS-positive droplets, immature germ cells, and multinucleate giant cells in the lumen. TPCP produced a decrease in sperm density but no other testicular effects at a dose level of 100 mg/kg per day.

To study the time course of the TOCP-induced testicular lesion in F-344 rats, the onset of possible changes in sperm numbers and production, serum hormones, and various enzyme activities was followed. Rats were administered TOCP daily (150 mg/kg) for periods of 3, 7, 10, 14, or 21 days, while vehicle-treated animals served as controls. Both sperm motility and the number of sperm per mg of cauda epididymis were lower in treated animals by day 10. The ratio of testicular to body weight was significantly decreased only those rats treated for 21 days. Testicular neurotoxic esterase and nonspecific esterase activities were also inhibited, while β-glucuronidase activity was not affected. Luteinizing and follicle-stimulating hormone levels were normal, as were both serum and interstitial fluid testosterone concentrations. Sertoli cell fluid secretion, as measured by testis weight increase after efferent duct ligation, showed no significant changes. Other organs (spleen, liver, kidney, pancreas, small intestine, and adrenal and pituitary glands) revealed no overt signs of pathology as observed by light microscopy in animals treated for 21 days. A separate group of animals was treated for 21 days and subsequently examined after 98 days of observation (two cycles of the rat seminiferous epithelium). Normal spermatogenesis did not return, indicating that the toxicity was irreversible at the dose used. The effects noted in these studies further define the testicular lesion produced by TOCP, and show that a dose level 150 mg/kg per day for 21 days produces irreversible testicular toxicity (Somkuti et al., 1987b).

The testicular effects of TOCP have also been studied in the rooster, 100 mg/kg per day being administered orally to ten adult leghorn roosters for 18 consecutive days. By days 7-10 of the study, TOCP-treated birds exhibited limb paralysis characteristic of OPIDN. Enzyme analyses at the end of the study revealed significant inhibition of neurotoxic esterase (NTE) activity in both brain and testis and a slight decrease in brain acetylcholinesterase (AChE) activity. Sperm motility was shown to be greatly decreased. In addition, sections of formalin-fixed, methacrylate-embedded testes from TOCP-treated birds showed vacuolation and disorganization in the seminiferous epithelium. The marginal body weight decrease (17%) in treated animals was not considered to contribute to the testicular toxicity induced by TOCP (Somkuti et al., 1987c).

TCP (isomer mixture not known) was tested in sperm morphology and vaginal cytology examination (SMVCE) studies in groups of 25 male and female Fisher-344 rats and 25 male and female Swiss CD-1 mice. Treatment lasted for 13 weeks at dose levels of 1700, 3300, or 6600 mg/kg diet or 50, 100, 200, 400, or 800 mg/kg body weight via gavage in the rats, and 500, 1000, or 2100 mg/kg diet or 50, 100, or 200 mg/kg body weight via gavage in the mice. Effects were seen in treated animals, but no details were given (Morrissey et al., 1988a). In a follow-up continuous breeding reproduction study in Swiss CD-1 mice, male and female fertility was reduced (Morrissey et al., 1988b).

Carlton et al. (1987) examined the reproductive effects of TCP (mixed isomers; < 9% TOCP). Male Long-Evans rats received 0, 100, or 200 mg/kg and females received 0, 200, or 400 mg/kg in corn oil by gavage. The low-dose males were mated with low-dose females, and the high-dose males with high-dose females. Males were dosed for 56 days and females for 14 days prior to breeding and throughout the breeding period, gestation, and lactation. Sperm concentration, motility, and progressive movement were decreased in the high-dose males, and there was a dose-dependent increase in abnormal sperm morphology. The number of females delivering live young was severely reduced by TCP exposure. Litter size and pup viability were decreased in the high-dose group, but pup body weight and developmental landmarks were unaffected by TCP exposure. Histological changes were observed in the testes and epididymides of treated males (i.e. necrosis, degeneration, early sperm granulomas in seminiferous tubules, and epididymal hypospermia) and in the ovaries of treated females (i.e. diffuse vacuolar cytoplasmic alteration of interstitial cells and increased numbers of follicles and corpora lutea).

In a continuous breeding study, Chapin et al. (1988) fed Swiss CD-1 mice a diet containing TCP (mixed o-, m-, p-isomers) at a level of 0, 0.5, 1, or 2 g/kg diet over 98 days. Although the fertility index was not changed in the high-dose groups, the number of litters per pair decreased in a dose-dependent fashion, and the proportion of pups born alive and their weight were significantly decreased. Histopathological examination of the parental generation revealed dose-related seminiferous tubule atrophy and decreased testis and epididymal weights in the high-dose males, but the female reproductive tract showed no histopathological changes. A crossover mating trial revealed impaired fertility in both males and females exposed to 2 g/kg diet.

8.6 Mutagenicity and carcinogenicity

Haworth et al. (1983) reported a negative result with TCP (substitution pattern not known) in the Salmonella mutagenicity test.

8.7 Neurotoxicity

8.7.1 Experimental neuropathology

TOCP first gained notoriety as the culpable neurotoxic agent of the "Ginger Jake" epidemic (Jeter, 1930; Goodale & Humphreys, 1931; Vonderahe, 1931). Since then several experimental studies have modelled TOCP neuropathy in various species.

Smith & Lillie (1931) produced delayed paralysis in various animals, including dogs, monkeys, cats, and chickens. These animal models were used to describe the functional and morphological features of TOCP neuropathy. After a delay period of 2-3 weeks following exposure to single or multiple doses, paralysis of the hindlegs developed in these species in response to the TOCP. Neuropathologically, degeneration was confined to the spinal cord and peripheral nerve fibres. These changes were essentially similar to the lesions reported in human victims of the "Ginger Jake" epidemic.

The delayed neuropathy associated with TOCP and other organophosphorous compounds has been termed OPIDN.

Cavanagh (1964) showed that, in cats and chickens, degeneration affected the long fibres in the spinal cord and peripheral nerves; moreover, in cats, fibre diameter seemed to be important in determining the onset and severity of the peripheral nerve lesions. He categorized this delayed neuropathy as a "dying-back". Thus, the lesions of the ascending tracts are seen in the cervical region, especially in the dorsal columns (i.e. fasciculus gracilis), while those of the descending tracts are seen in the thoracic and lumbosacral regions of the spinal cord. He also showed that the peripheral and central nerve lesions were the result of axonal degeneration and did not reflect primary demyelination. The affected nerve axons degenerate in a "dying-back" fashion towards the cell body, i.e. axonal degeneration begins at the most distal portion of the axon and proceeds towards the cell body.

Prineas (1969) described axons with accumulated tubulovesicular membranes within myelinated motor nerve terminals in the foot muscles of cats dosed with TOCP. Similar early axoplasmic changes have been noted in TOCP-treated chickens (Bischoff, 1967, 1970; Spoerri & Glees, 1979, 1980).

At the ultrastructural level, the nerve endings are characterized by a marked proliferation and distension of vesicular elements of the endoplasmic reticulum and a coinciding disintegration of the filamentous and tubular organelles (Bischoff, 1977). Bischoff also noted that the presynaptic nerve terminals and *boutons terminaux* appeared particularly sensitive in TOCP intoxication.

OPIDN has also been produced in rats dosed with TOCP (Veronesi, 1984). The pathological changes developed in the absence of discernable ataxia after a 2-week exposure to acute and multiple oral doses of TOCP (> 840 mg/kg). In the rat, dorsal column degeneration of the cervical cord and a selective vulnerability of large diameter tibial branches supported a "dying-back" neuropathy, as in other species.

OPIDN has also been described in European ferrets (*Mustela putorius furo*) administered a single oral or dermal dose of 250, 500, or 1000 mg TOCP/kg body weight. Five animals per group were sacrificed 48 h after dosing, the others being observed for another 54 days. All ferrets dosed dermally with 1000 mg/kg developed neurological signs ranging from ataxia to partial paralysis. Dermal doses of 250 and 500 mg/kg produced variable degrees of hind limb weakness and ataxia. Of the animals dosed orally, only those treated with 1000 mg/kg showed neurological signs, which did not progress beyond mild ataxia. Slight axonal degeneration was noted in the dorsolateral part of the lateral funicilus and in the fasciculus gracilis of spinal cords in ferrets receiving a dermal dose of 1000 mg/kg. Whole brain neurotoxic esterase activity was maximally inhibited (46%) at this dose level. The study demonstrated that in the ferret dermal exposure was more effective than oral exposure at the same dose level (Stumpf et al., 1989).

8.7.2 Neurochemistry

Johnson (1969) found that approximately 6% of brain esterases are not affected by non-neuropathic compounds but are specifically inhibited, irreversibly, by neuropathic ones, such as TOCP. Johnson used the term "neurotoxic esterase" (NTE) and proposed that NTE is the primary target of the organophosphorus esters causing OPIDN (Johnson, 1975a,b). There appears to be a strong correlation in the chicken between NTE inhibition above 70% shortly after exposure and subsequent neuropathy for a large number of tested organophosphorous compounds (Johnson, 1974).

Padilla & Veronesi (1985) demonstrated the relevance of NTE to the rodent model of OPIDN by exposing Long-Evans rats to single doses of TOCP ranging from 290 to 3480 mg/kg. High NTE inhibition in the spinal cord (> 72%) and the brain (> 66%) produced severe spinal cord damage in over 90% of exposed rats, indicating that NTE depression could predict OPIDN damage in rats acutely exposed to organophosphates.

Concerning the role of lipids in TOCP neuropathy, Morazain & Rosenberg (1970) showed that there was a rise of 25-50% in the cholesterol level in the sciatic nerve and a 50% decrease in its triglyceride content in chickens orally dosed with 1 ml TOCP/kg. Phospholipids, diglycerides, cholesterol esters, proteolipids, and tissue phospholipases were not affected by TOCP. The possible involvement of lipids in the production of TOCP-induced delayed neuropathy has not yet been resolved.

Other toxic effects have been demonstrated. Brown & Sharma (1975) found that neural membrane ATPases were inhibited by organophosphates. Cohen & Murphy (1970) reported that TOCP potentiates the anticholinesterase action of malathion by 29-fold in mice, 17-fold in quail, and 11-fold in sunfish.

8.7.3 Interspecies sensitivity and variability to OPIDN

Certain animal species (e.g., cats, dogs, cows, and chickens) are susceptible to OPIDN-related paralysis, whereas others (e.g., rats and mice) are less susceptible to the ataxia but very susceptible to the pathological changes. Species susceptibility to delayed neurotoxicity induced by TOCP shows an inverse correlation with the rate of metabolic conversion to the neurotoxic metabolite (see section 7). Because of its high susceptibility to ataxia, the adult chicken has been used as an experimental model to study OPIDN.

TOCP is metabolized to the more potent neurotoxic agent, saligenin cyclic *o*-tolyl phosphate, which is at least five times more neurotoxic than TOCP after oral administration to chickens: a metabolite level of 40 mg/kg caused ataxia equivalent to that resulting from 200 mg TOCP/kg (Bleiberg & Johnson, 1965).

It has been shown that a single oral dose of TOCP in the range of 58 to 580 mg/kg (i.e. 0.05-0.5 ml/kg) induces mild to severe paralysis in the hen *(Gallus domesticus)* (Cavanagh, 1954; Hine et al., 1956).

Johannsen et al. (1977) showed that chickens administered cumulative doses of TCP (60 000 mg/kg) or TOCP (1500 mg/kg) developed both the ataxic and neuropathological symptoms of OPIDN.

In 90-day studies in hens, obvious functional and morphological neuropathological changes were found at daily oral dose levels of 5 to 20 mg TOCP/kg body weight, but not at lower dosages (Smith et al., 1932; Prentice & Majeed, 1983; Roberts et al., 1983).

In a subchronic feeding study, Haggerty et al. (1986) exposed rats to TCP (isomeric substitution pattern not known) for 13 weeks at dose levels of 0, 900, 1700, 3300,

6600, or 13 000 mg/kg. Decreased hindlimb grip strength was observed in male rats (at 13 000 mg/kg), but not in female rats. In mice exposed to 0, 250, 500, 1000, 2100, or 4200 mg/kg, decrements in both fore- and hindlimb grip strength were observed in males (at 4200 mg/kg) and females (at 2100 and 4200 mg/kg). A reduction of body weight was seen both in rats and mice at the two highest dose levels. Preliminary histopathological diagnosis indicated demyelination and axonal degeneration of the sciatic nerve in male and female mice only.

Freeman et al. (1988) tested the neurotoxicity of TCP (substitution pattern not known) to F-344 rats in a short-term study. After 13 weeks of dosing with TCP in the feed, hindlimb grip strength decreased in male rats at 300 and 600 mg/kg, but not in female rats. Serum cholinesterase was reduced at 300 and 600 mg/kg in both males and females. All effects observed with 600 mg/kg were apparently reversible during the recovery period.

In a 13-week short-term feeding study, Irwin et al. (1987) exposed F-344 rats to TCP (0, 75, 150, 300, or 600 mg/kg; substitution pattern not known), while $B6C3F_1$ mice receive 0, 60, 125, or 250 mg/kg. After 13 weeks of dosing, forelimb grip strength was unaffected by TCP in both mice and rats. Hindlimb grip strength decreased in male rats (at 300 and 600 mg/kg) but not in female rats. In mice, decrements in hindlimb grip strength were observed in males (250 mg/kg) and females (125 and 250 mg/kg). Serum cholinesterase levels showed a dose-dependent reduction in both rats and mice. TCP had no effect on food consumption in either species. All groups exhibited normal body weight values.

Factors such as age, sex, and strain figure prominently in the expression of OPIDN. The young of most species are non-susceptible to TOCP-induced delayed neuropathy (Johnson & Barnes, 1970), which could be due to poor absorption of TOCP. However, recent experiments (Olson & Bursian, 1988) have suggested that factors (e.g., route of administration) other than absorption are more critical to this lack of susceptibility.

Strain differences in the susceptibility of rats to TOCP-delayed neuropathy have been reported. Although OPIDN can be readily produced in Long-Evans and Sprague-Dawley rats after acute oral doses of > 840 mg/kg (Veronesi & Abou-Donia, 1982; Veronesi, 1984), repeated doses of TOCP (10-100 mg/kg) failed to produce neuropathy in Fischer-344 rats (Somkuti et al., 1988). Variations in TOCP inhibition of brain AChE and NTE have also been reported in these three strains (Carrington & Abou-Donia, 1988).

8.7.4 Neurophysiology

Robertson et al. (1987) investigated electrophysiological changes in the adult hen following single oral doses of 30 or 750 mg TOCP/kg. At the higher dose, the birds demonstrated clinical signs of toxicity 12 days after dosing that included gait abnormalities, which became progressively more severe and in some cases led to complete ataxia. Lymphocyte NTE was inhibited by more than 70%. The lower dose produced no clinical signs of toxicity and only 54% lymphocyte NTE inhibition. Both treated groups displayed significant action potential disruption in both the tibial and sciatic nerves that resulted in decreased refractoriness in the tibial nerve, increased refractoriness in the sciatic nerve, and elevated strength duration threshold for both nerves.

Electrophysiological changes have been investigated in cats following single dermal doses ranging from 250 to 2000 mg TOCP/kg and 90-day dermal administration of 1 to 100 mg/kg (Abou-Donia et al., 1986). In contrast to the hen, the clinical signs of TOCP neurotoxicity appear in the cat 21-26 days before the electrophysiological effects on the gastrocnemius muscle. Recovery of the cat from delayed neurotoxicity symptoms was more marked than that of the hen. No effects on peripheral nerve transmission or on neuromuscular junction functioning were seen in the cat.

9. EFFECTS ON HUMANS

Summary

There have been many reported cases of human poisoning, mostly from accidental or irresponsible contamination of foodstuffs. Occupational poisoning, usually resulting from dermal exposure, has also been reported. The ortho isomer of TCP is the responsible toxic agent.

Though short-term symptoms of ingestion might involve vomiting, abdominal pain, and diarrhoea, characteristically delayed, longer-term symptoms are neurological, frequently leading to paralysis and pyramidal signs (spasticity, etc.).

There is considerable variation in the sensitivity of individuals to TOCP; severe symptoms were reported with a TOCP dose of 0.15 g in one individual, while others were unaffected by 1 to 2 g. There is also considerable variation in the rate of recovery from poisoning, some patients recovering completely and others still severely affected years later, after apparently similar exposure.

First-aid treatment involves the induction of vomiting or pumping of the stomach. The patient should be hospitalized as soon as possible. Atropine or 2-PAM may be used as an effective antidotal treatment against cholinergic symptoms. Long-term, antispastic drugs may be useful, though physical rehabilitation is the cardinal therapy.

9.1 Historical background

Of the tricresyl phosphate isomers, the ortho (TOCP) is by far the most toxic and alone gives rise to the major neurotoxicity in man. It is considered that the toxicity of the commercial products depends on the concentration of the ortho isomer, but the mixed o-cresyl esters in these products are also toxic and contribute to the neurotoxic action.

It is well known that TOCP produces delayed effects on the central and peripheral nervous systems. TOCP poisoning has occurred throughout the world (Inoue et al., 1988); the major outbreaks are indicated in Table 14.

Table 14. Major outbreaks of TOCP poisoning

Year	Place	Number of cases	Vehicles of TOCP	Reference
1898	France	6	phospho-creosote	Lorot (1899)
1900-1928	Europe	43	phospho-creosote (15% TOCP)	Roger & Recordier (1934)
1930-1931	USA	50 000	ginger extract	Morgan (1982)
1931	Europe	several hundred	Apiol pill	Susser & Stein (1957)
1938	South Africa	68	cooking oil	Sampson (1942)
1940	Switzerland	80	cooking oil	Walthard (1945)
1941-1945	Germany	more than 200	cooking oil	Mertens (1948)
1945	England	17	cooking oil	Hotston (1946)
1955	South Africa	11	water or solvent	Susser & Stein (1957)
1957	Morocco	about 10 000	cooking oil	Smith & Spalding (1959)
1960	India	58	solid food	Vora et al. (1962)
1962	India	more than 400	flour	Chaudhuri (1965)
1967	Fiji	56	flour	Sorokin (1969)
1980	Romania	12	alcohol	Vasilescu & Florescu (1980)
1981	Sri Lanka	more than 20	cooking oil (0.56%)	Senanayake & Jeyaratnam (1981)

In 1899, Lorot initially reported six cases of polyneuropathy out of 41 cases of pulmonary tuberculosis treated with phospho-creosote. Later it was shown that the phospho-creosote contained 15% TOCP. In the next 30 years, 43 additional cases caused by the drug were reported in various parts of continental Europe (Roger & Recordier, 1934).

In the spring of 1930, in mid-western and south-western USA, an outbreak of paralysis characterized by bilateral foot- and wrist-drop appeared suddenly (Burley, 1930; Merritt & Moor, 1930). Ultimately 50 000 people were poisoned by a popular substitute for alcohol called "Ginger Jake" (Morgan, 1982). Smith et al. (1930) proved that the adulterated beverage contained about 2% TOCP and that this caused the paralysis.

In 1931, several hundred women in Europe (especially in Germany, The Netherlands, Yugoslavia, and France) were poisoned by the TOCP contained in Apiol pills and taken as an abortifacient (Roger & Recordier, 1934; Susser & Stein, 1957). The TOCP was presumably included as an additional stimulus to abortion (Ter Braak & Carrillo, 1932).

An outbreak involving 11 people occurred in Durban, South Africa, in 1955 (Susser & Stein, 1957). From the epidemiological survey it was suggested that water, as well as solvents, may have been a vehicle for the TOCP.

In 1959, about 10 000 Moroccan people were intoxicated by TOCP: jet engine oil had been illegally mixed into their cooking oil (Smith & Spalding, 1959; Svennilson, 1960). Accidental poisoning by TOCP contamination of solid food occurred in 1960 in Bombay, India, where 58 victims were recorded (Vora et al., 1962).

During the period April-June, 1962, more than 400 cases of paralysis occurred in the Malda district in India. The cause of this disease proved to be the consumption of flour contaminated with TOCP (Chaudhuri, 1965).

In 1967, similar poisoning was recorded in Fiji, where 56 people showed neuropathy (Sorokin, 1969). The cause was stated to be contamination of dry sharps flour by TOCP through the sacking material.

Vasilescu & Florescu (1980) in Romania reported 12 patients with toxic neuropathy following accidental ingestion of alcohol contaminated by TOCP.

An outbreak of acute polyneuropathy in over 20 young females occurred in Sri Lanka during 1977-1978 (Senanayake & Jeyaratnam, 1981). The cause of the neuropathy was traced to TCP found as a contaminant in a special cooking oil (gingili oil). Contamination probably occurred during transport of the oil in containers previously used for storing mineral oils.

9.2 Occupational exposure

Gartner & Elsaesser (1943) reported the case of a worker who developed pyramidal signs after exposure to TOCP for two years in a German chemical plant. In this case, percutaneous absorption was considered to be the main route of exposure.

In 1944, three cases of toxic polyneuropathy among workers who had worked for six to eight months in a plant manufacturing TCP in England were reported (Hunter et al., 1944). Skin penetration and inhalation were thought to be the main causes of the occupational poisoning.

Parnitzke (1946) reported a case of TOCP poisoning after 3 years of exposure in a German plant and stated that TOCP had been absorbed through the skin and presumably the gastrointestinal tract.

Since 1958, a high prevalence of polyneuropathy among shoe factory workers has been reported in Italy. The cause has been attributed to TCP (Cavalleri & Cosi, 1978). Although this is possible, this cause-effect relationship has not so far been based on unquestionable evidence. This polyneuropathy might have various aetiological factors (including *n*-hexane) or be produced by a combination of them (Leveque, 1983).

9.3 Clinical features

Goldstein et al. (1988) reported a case of severe intoxication in a 4-year-old child following ingestion of a lubricant containing TCP (substitution pattern not known). The clinical findings were acute gastrointestinal symptoms, delayed cholinergic crisis, and neurological toxicity.

The severity of signs and symptoms after poisoning with TOCP seems not always to be proportional to the dosage (Staehelin, 1941). In a Swiss Army outbreak of poisoning among more than 80 young men, toxic symptoms appeared in once case after eating food containing only 0.15 g TOCP. Severe neurological disturbance developed in three men from the intake of 0.5 to 0.7 g, whereas in two other cases the intake of 1.5-2 g did not lead to any symptoms. This leads to the conclusion that individual susceptibility varies greatly (Staehelin, 1941).

In general, the signs and symptoms of TOCP poisoning are distinctive, whereas the symptomatology varies somewhat according to whether a single relatively large dose or small cumulative doses are taken. In the former case, the initial symptoms are gastrointestinal, ranging from slight to severe nausea and vomiting, sometimes accompanied by abdominal pain and diarrhoea. Among these symptoms, vomiting is most frequently observed (Staehelin, 1941). These symptoms are usually transient, lasting from a few hours to a few days (Walthard, 1945; Susser & Stein, 1957).

In cases of chronic low level exposure, the above symptoms may not be present and the major symptoms are neurological (Parnitzke, 1946). The clinical features of acute exposure to TOCP were described by Staehelin (1941). A latent period of 3-28 days is observed after acute exposure, and clear "delayed neurotoxicity" then gradually appears. The initial neurological symptoms are sharp, cramp-like pains in the calves, and some numbness and tingling in the feet and sometimes the hands. Within a few hours or a day or two at most, these pains are followed by increasing weakness of the lower limbs, and soon the patient becomes unsteady and then unable to maintain balance. The cramp-like pains may cease with the onset of weakness, or persist for some days. One or two weeks after the onset in the lower limbs and while paralysis may still be progressing, the weakness spreads to the hands. While some patients show complete wrist drop and total loss of power in the hands, sometimes with weakness up to the elbows, the predominant neurological abnormalities are observed in the lower limbs. Bilateral foot drop with complete loss of power from the ankle down is a common finding. Depending on the severity of the affection, the patient may have weakness in the knees, less at the hips, and, only in the most severe cases, weakness of the trunk. About three weeks or more after the onset of paralysis, a most striking and rapid wasting may be observed. While the small muscles of the feet, calves, the anterior tibials, and the thighs do not escape this wasting, in so far as they are involved in the disease, the change is most obvious in the small muscles.

In the initial stage, the ankle jerks are absent and knee reflexes may be normal or occasionally depressed. Plantar reflexes are unobtainable. Mild cases do not show any upper motor neuron signs. On the other hand, in the more severe cases, even at the early stage, knee jerks may be exaggerated, presaging the development of upper motor neuron involvement. In general, upper motor neuron signs, e.g., pyramidal signs, gradually become evident at about the third week or later. Knee jerks become exaggerated and so also may the biceps, triceps, and supinator jerks (Cavanagh, 1964). A finger flexor reflex appears for the first time or increases (Susser & Stein, 1957). As the pyramidal tract lesion becomes evident, involuntary flexor withdrawal of the whole limbs follows gentle plantar stimulation. Babinski responses are observed much later. Muscle tonus of the limbs gradually increases. In severe cases, the signs of upper motor involvement are delayed, probably masked by the gross flaccid muscle weakness.

Several authors state emphatically that sensory disturbances do not occur. Sampson (1942) reported sensory disturbances although these were admittedly unobtrusive, in contrast to motor dysfunction. Reports of muscle and peripheral nerve tenderness are fairly frequent. If the sensory disturbances are observed, there is hypoesthesia with loss of pin-prick and temperature sense; the ability to detect vibration is sometimes affected distally. The sensory disturbances vary in extent from merely the soles of the feet to the whole of the limbs.

Usually cranial nerves are not involved. In general, mental signs are rare, but transient euphoria and confusion have been observed in the early stage (Schwab, 1948).

9.4 Prognosis

Following exposure, muscle weakness progresses over several weeks, sometimes even months. Sensory changes often begin to regress during the early weeks, the rapidity depending on the severity of the case, and then muscle strength gradually returns in patients who are only mildly affected. Improvement begins with the return of sensation, then muscle strength in the hands, and eventually strength in the lower limbs. In cases of pyramidal signs, recovery is generally poor. Zeligs (1938), reporting eight years after the 1930 mid-western and south-western USA outbreak, surveyed the records of 316 patients. He was able to follow up 60 patients, all of whom were disabled and still in institutions. Aring (1942) examined more than 100 patients in the 10 years following this outbreak and they appeared still to be affected. Morgan & Penovich (1978) followed up 11 survivors in the 47 years after the same outbreak; the principal findings were the spasticity and abnormal reflexes of an upper motor neuron syndrome.

Of the 80 patients in the 1940 Swiss army accident, 14 were quite well after five years, 15 were totally incapacitated, and 38 showed spasticity (Walthard 1945).

In the 1938 Durban outbreak, all the patients showed some symptoms of the disease 18 years later (Susser & Stein, 1957). The mildest case had slight weakness at the ankle, while the most severely affected had foot drop, muscle atrophy, and pyramidal signs (spasticity, ankle clonus, and positive Babinski sign).

The residual signs and symptoms are mainly confined to the lower limbs. They consist of weakness and muscle atrophy of varying degrees in the feet and the small muscles of the hand. Disability is principally related to the pyramidal signs with resultant spasticity of the lower extremities.

9.5 Neurophysiological investigations

There have been very few electrophysiological studies on human TOCP poisoning. Svennilson (1960) reported an electromyographic study on 65 patients in the 1959 Morocco poisoning. These cases showed varying degrees of denervation and polyphasic abnormal potentials in the paralysed muscles. The clinically healthy proximal groups of muscles also showed marked polyphasic action potentials but not denervation.

Vasilescu & Florescu (1980) reported detailed studies on 12 patients from the 1980 accident in Romania. They observed > 50% decrease in the muscle evoked potential amplitude, fibrillation potentials in the same muscles at rest, and decreased motor nerve conduction velocity.

In neurophysiological studies by Senanayake (1981) in Sri Lanka, the main findings were widespread neurotoxic patterns and prolongation of terminal latencies with relatively mild slowing of motor nerve conduction velocities. These studies confirmed the evidence of axonal degeneration.

9.6 Pathological investigations

Numerous pathological studies have been made on biopsy or autopsy samples since the Jamaica ginger accidents in 1930. In 1930, Goldfain described some changes observed in the peripheral nerves and spinal cord, quoting Jeter's autopsy report. Histopathological investigations by Goodale & Humphreys (1931) indicated degeneration of myelin sheaths and axis cylinders in the radial, sciatic, and tibial nerves in all cases examined. Vonderahe (1931) found marked degenerative changes in the anterior horn cells, characterized by swelling, central chromatolysis (disappearance of the Nissl substance), excentric nuclei, and shrinking of the cells. The pathological studies also revealed degeneration in the radial and anterior tibial nerves, and degenerative changes in the anterior roots. There were no pathological signs of inflammation. According to Smith & Lillie (1931), the paralysis due to Jamaica ginger was essentially a degeneration of the myelin sheaths of the peripheral nerves, with a variable amount of relatively moderate central degenerative changes affecting the anterior horn cells throughout the spinal cord, but more often in the lumber and cervical regions. In the 1938 Durban (South Africa) epidemic, Sampson (1942) also reported that degeneration of the anterior horn cells occurred in some instances and that peripheral nerves showed axonal degeneration. Aring (1942) described degeneration in the posterior and lateral columns in later investigations of survivors of the outbreak, thereby confirming the origin of some of the spinal symptoms. It is noteworthy that the latter changes were evident in the lumbar region, while the dorsal column changes, in which only the fasciculus gracilis was involved, concerned the cervical region.

Muscle biopsy studies of patients from the 1959 Moroccan poisoning showed a moderate degree of muscle atrophy and a slight increase of the muscle fibre nuclei. Spherical axonal swelling and terminal knobs were noted as a sign of peripheral nerve degeneration in muscles (Svennilson, 1960). Similar changes were also noted in the poisoning cases reported in Malda, India, in 1962 (Chaudhuri et al., 1962). These effects on muscle suggest denervation.

9.7 Laboratory investigations

Little information has been obtained from laboratory examinations of exposed humans. There is no significant change in the urine or blood (Sampson, 1942; Senanayake & Jeyaratnam, 1981), but the cerebrospinal fluid may show an increase in protein concentration, with or without pleocytosis (Sampson, 1942; Mertens, 1948; Susser & Stein, 1957).

Vora et al. (1962) measured blood cholinesterase levels in patients admitted to hospitals during the 1960 Bombay poisoning and demonstrated that plasma cholinesterase was increased one month after exposure. The erythrocyte cholinesterase level was considerably diminished (50%) quite early after the onset of symptoms. Both plasma and erythrocyte cholinesterase activities returned to normal within about 3 months.

In a factory manufacturing TAP, about half of the workers examined showed significant decreases in pseudocholinesterase (cholinesterase other than AChE), and many of them had minor signs and symptoms (Tabershaw et al., 1957).

Morgan & Hughes (1981) also investigated cholinesterase activity in workers in a plant manufacturing TAP plasticizers. They found that plasma cholinesterase estimation in workers exposed to TAP compounds cannot be used as a sensitive barometer of organic phosphate absorption; thus routine regular estimations serve no useful purpose. Its value is chiefly as a screening method at the pre-employment medical examination to exclude personnel at risk, as a baseline in the event of massive exposure, and as a means of diagnosis in cases of CNS disease that simulate TAP poisoning.

9.8 Treatment

In the event of skin contact with TCP, contaminated clothing should be rapidly removed and affected body areas copiously irrigated with water. The ingestion of food or drink contaminated with TOCP should be treated by inducing vomiting, unless the patient is unconscious. Atropine or pralidoxine (2-PAM) chloride may be required to counteract cholinergic effects. No specific antidote is available.

Medical therapy should begin as soon as possible, even though the results of medical therapeutic measures have been disappointing. B-complex vitamins and corticosteroids may protect nervous tissue against further involvement (Geoffroy et al., 1960). The cardinal therapy is physical rehabilitation. Administration of anti-spastic drugs may be required.

10. EVALUATION OF HUMAN HEALTH RISKS AND EFFECTS ON THE ENVIRONMENT

10.1 Evaluation of human health risks

Human poisoning involving the accidental ingestion of tri-*o*-cresyl phosphate (TOCP) or occupational exposure of workers has frequently been reported. The likely route of occupational exposure is cutaneous absorption. The neurotoxic symptoms involve initial inhibition of cholinesterases and subsequent delayed neuropathy characterized by severe paralysis.

Because of considerable variation among individuals in sensitivity to TOCP, it is not possible to establish a safe level of exposure. Symptoms have been reported from the ingestion of 0.15 g of an isomeric mixture with a low proportion of TOCP; the minimum effective dose of the ortho isomer is, therefore, much lower than this. Animal studies show considerable variation among species in the response to TOCP, and humans appear to be particularly sensitive.

Irritant and allergic dermatitis have been reported.

Both the pure ortho isomer and isomeric mixtures containing TOCP are, therefore, considered major hazards to human health.

There is no safe level for ingestion. Exposure to the compound through dermal contact or inhalation should be minimized.

10.1.1 Exposure levels

Exposure of the general population to tricresyl phosphate (TCP) through various environmental media, including drinking-water, can be regarded as minimal. TCP has been detected at relatively higher concentrations in urban air than in air collected at production sites, although the levels are usually low. TCP was not detected in human adipose tissue samples in a survey conducted in the USA. There have been many cases of accidental human poisoning through the ingestion of medicines, food, flour, cooking oil, and beverages contaminated with hydraulic fluid or lubricant oil containing TCP produced from "cresylic acid". The toxic symptoms can be observed after ingestion of only 0.15 g of tri-*o*-cresyl phosphate, a component of TCP from cresylic acid. The contamination has usually happened when empty barrels or drums, previously used for hydraulic fluid or lubricating oil storage, have been reused.

10.1.2 Toxic effects

Accidental human exposure to a single large dose results in gastrointestinal disturbance varying from slight to severe nausea and vomiting, accompanied by abdominal pain and diarrhoea. In the case of exposure to small cumulative doses, "delayed neurotoxicity" gradually proceeds after a latent period of 3-28 days. In most cases, the muscle weakness changes rapidly to a striking paralysis of the lower limbs, with or without an involvement of the hands. In severe cases, pyramidal signs gradually become evident. Some neurophysiological studies indicate widespread neurotoxic patterns and prolongation of terminal latencies with relatively small decreases of motor nerve conduction velocities. This confirms the evidence of axonal degeneration, which is the main feature observed in pathological investigations.

The neurotoxic metabolite of TCP has been identified as saligenin cyclic *o*-tolyl phosphate, which is derived from *o*-hydroxymethyl metabolites. Thus, it seems that at least one *o*-tolyl group among the three phenolic moieties of TCP is necessary to induce neurotoxic effects. TCP produced from synthetic cresol, which contains less than 0.1% of *o*-cresol, is therefore not neurotoxic.

Subchronic animal studies on TCP derived from synthetic cresol indicate that the target organs are liver and kidney, but this was not confirmed in the case of human intoxication. No adequate data are available on mutagenicity and carcinogenicity. TCP is not toxic to chick embryos.

10.2 Evaluation of effects on the environment

The measurement of environmental concentrations of TCP in water has shown only low levels of contamination. This reflects the low water solubility and ready degradability of the compound. Since the acute toxicity of TCP to aquatic organisms is also low, it is unlikely that it poses a threat to such organisms.

As a consequence of the physico-chemical properties of TCP, there is a high potential for bioaccumulation. However, this does not occur in practice, owing to low concentrations of TOCP in the environment and living organisms and to its rapid degradation.

TCP bound to sediment accumulates in the environment, and levels measured in river, estuarine, and marine sediments have been high. Since there is no information either on the bioavailability of these residues to burrowing or bottom-living organisms or on their hazards, the possibility of effects on such species cannot be discounted.

TCP spillage leads to hazard for the local environment.

10.2.1 Exposure levels

TCP is found in air, surface water, soil, sediment, and aquatic organisms near heavily industrialized areas, although concentrations are usually low. Owing to the high biodegradation rate of TCP in aqueous environment, it is not considered to affect aquatic organisms adversely. One report showed an extremely high concentration of total triaryl phosphate (26.55 g/kg) in a soil sample obtained from a production plant yard. This suggests the need for land waste disposal.

10.2.2 Toxic effects

Freshwater algae are relatively sensitive to TCP, the 50% growth inhibitory concentration ranging from 1.5 to 5.0 mg/litre. Among fish species, the rainbow trout is adversely affected by TCP concentrations below 1 mg/litre (0.3-0.9 mg/litre), with sign of chronic poisoning, but the tidewater silverside is more resistant (LC_{50} is 8700 mg/litre). TCP does not inhibit cholinesterase activity in fish or frogs, but it has a synergistic effect on organophosphorus insecticide activity.

11. RECOMMENDATIONS

When tri-substituted cresols are used in the synthesis and manufacture of other compounds, the purified meta and para isomers should be used in order to avoid the accidental synthesis of ortho-substituted products.

REFERENCES

ABOU-DONIA, M.B., VARGA, P.A., GRAHAM, D.G., & KINNES, C.G. (1980) Delayed neurotoxicity of a single dermal administration of o-ethyl o-4-nitrophenyl phenylphosphonothioate and tri-o-cresyl phosphate to cats. *Toxicol. Lett.*, 1 (Special issue): 141.

ABOU-DONIA, M.B., TROFATTER, L.P., GRAHAM, D.G., & LAPADULA, D.M. (1986) Electromyographic, neuropathologic and functional correlates in the cat as the result of tri-o-cresyl phosphate delayed neurotoxicity. *Toxicol. appl. Pharmacol.*, 83(1): 126-141.

ADEMA, D.M.M., KUIPER, J., HANSTVEIT, A.O., & CANTON, H.H. (1983) Consecutive system of tests for assessment of the effects of chemical agents in the aquatic environment. In: *Pesticide chemistry-Human welfare and the environment. Proceedings of the Fifth International Congress on Pesticide Chemistry, Kyoto, Japan, 29 August - 4 September, 1982,* Oxford, New York, Pergamon Press, Vol. 3, pp. 537-544.

ALDRIDGE, W.N. (1954) Tricresyl phosphates and cholinesterase. *Biochem. J.,* 56: 185-189.

ALDRIDGE, W.N. & BARNES, J.M. (1961) Neurotoxic and biochemical properties of some triaryl phosphates. *Biochem. Pharmacol.,* 6: 177-188.

ANONYMOUS (1986) *Handbook of commodity chemicals,* Tokyo, Kagaku-Kogyo-Nipposha.

ARING, C.D. (1942) The systemic nervous affinity of triorthocresyl phosphate (Jamaica ginger palsy). *Brain,* 65: 34-47.

ASSOCIATION OF THE PLASTICIZER INDUSTRY OF JAPAN (1976) *Safety of the plasticizer TCP, tricresyl phosphate,* Tokyo, Association of the Plasticizer Industry of Japan.

BALDRIDGE, H.D., JENDEN, D.J., KNIGHT, C.E., PREZIOSI, T.J., & TUREMAN, J.R. (1959) Toxicology of a triaryl phosphate Oil. III. Human exposure in operational use aboard ship. *Am Med. Assoc. Arch. ind. Hyg.,* 20: 258-261.

BARNARD, P.W.C., BUNTON, C.A., LLEWELLYN, D.R., VERNON, C.A., & WELCH, V.A. (1961) The reactions of organic phosphates. Part V. The hydrolysis of triphenyl and trimethyl phosphates. *J. Chem. Soc. (B),* 1961: 2670-2676.

BARNARD, P.W.C., BUNTON, C.A., KELLERMAN, D., MHALA, M.M., SILVER, B., VERNON, C.A., & WELCH, V.A. (1966) Reactions of organic phosphates. Part VI. The hydrolysis of aryl phosphates. *J. Chem. Soc. (B),* 1966: 227-235.

BARRETT, H., BUTLER, R., & WILSON, I.B. (1969) Evidence for a phosphorylenzyme intermediate in alkaline phosphatase catalyzed reactions. *Biochemistry,* 8: 1042-1047.

BECK, B.E., WOOD, C.D., & WHENHAM, G.R. (1977) Triaryl phosphate poisoning in cattle. *Vet. Pathol.,* 14: 128-137.

BELL, J.D. & RUSVINE, J.R. (1967) Synergism of organophosphates in *Musca domestica* and *Chrysomya putoria*. *Entomol. exp. appl.*, **10**: 263-269.

BHATTACHARYYA, J., BHATTACHARYYA, K., SENGUPTA, P.K., & GANGULY, S.K. (1974) Detection and estimation of tricresyl phosphate in mustard oil. *Forensic Sci.*, **3**: 263-270.

BISCHOFF, A. (1967) The ultrastructure of tri-ortho-cresyl phosphate poisoning. I. Studies on myelin and axonal alterations in the sciatic nerve. *Acta neuropathol.*, **9**: 158-174.

BISCHOFF, A. (1970) The ultrastructure of tri-ortho-cresyl phosphate poisoning in the chicken. II. Studies on spinal cord alterations. *Acta neuropathol.*, **15**: 142-155.

BISCHOFF, A. (1977) Tri-ortho-cresyl phosphate neurotoxicity. In: Roisin, L., Shiraki, H., & Grcevic, N., ed. *Neurotoxicology*, New York, Raven Press, pp. 431-441.

BLEIBERG, M.J. & JOHNSON, H. (1965) Effects of certain metabolically active drugs and oximes on tri-*o*-cresyl phosphate toxicity. *Toxicol. appl. Pharmacol.*, **7**: 227-235.

BLOOM, P.J. (1973) Application des chromatographies sur couche mince et gaz-liquide à l'analyse qualitative et quantitative des esters des acides phosphorique et phosphoreux. *J. Chromatogr.*, **75**: 261-269.

BOETHLING, R.S. & COOPER, J.C. (1985) Environmental fate and effects of triaryl and tri-alkyl/aryl phosphate esters. *Residue Rev.*, **94**: 49-99.

BONDY, H.F., FIELD, E.J., WORDEN, A.N., & HUGHES, J.P.W. A study of the acute toxicity of the tri-aryl phosphates used as plasticizers. (1960) *Br. J. ind. Med.*, **17**: 190-200.

BOWERS, W.D., PARSONS, M.L., CLEMENT, R.E., EICEMAN, G.A., & KARASEK, F.W. (1981) Trace impurities in solvents commonly used for gas chromatographic analysis of environmental samples. *J. Chromatogr.*, **206**: 279-288.

BROWN H.R. & SHARMA, R.P. (1975) Inhibition of neural membrane adenosine-triphosphatases by organophosphates. *Toxicol. appl. Pharmacol.*, **33**: 140.

BURLEY, B.T. (1930) The 1930 type of polyneuritis. *New Engl. J. Med.*, **202**: 1139-1142.

CARLTON, B.D., HASARAN, A.H., MEZZA, L.E., & SMITH, M.K. (1987) Examination of the reproductive effects of tricresyl phosphate administered to Long-Evans rats. *Toxicology*, **46**: 321-328.

CARRINGTON, C.D. & ABOU-DONIA, M.B. (1988) Variation between three strains of rat: inhibition of neurotoxic esterase and acetylcholinesterase by tri-o-cresyl phosphate. *J. Toxicol. environ. Health*, **25**(3): 259-268.

CASIDA, J.E. (1961) Specificity of substituted phenyl phosphorus compounds for esterase inhibition in mice. *Biochem. Pharmacol.*, **5**: 332-342.

CASIDA, J.E., ETO, M., & BARON R.L. (1961) Biological activity of a tri-o-cresyl phosphate metabolite. *Nature (Lond.)*, 191: 1396-1397.

CASTERLINE, J.L., Jr., KU, Y., & BARNETT, N.M. (1985) Uptake of tri-p-cresyl phosphate in soybean plants. *Bull. environ. Contam. Toxicol.*, 35: 209-212.

CAVALLERI, A. & COSI, V. (1978) Polyneuritis incidence in shoe factory workers: Case report and etiological considerations. *Arch. environ. Health*, 33: 192-197.

CAVANAGH, J.B. (1954) The toxic effects of tri-ortho-cresyl phosphate on the nervous system. An experimental study in hens. *J. Neurol. Neurosurg. Psychiatry*, 17: 163-172.

CAVANAGH, J.B. (1964) The significance of the "dying back" process in experimental and human neurological disease. *Int. Rev. exp. Pathol.*, 3: 219-267.

CHAPIN, R.E., GEORGE, J.D., & LAMB, J.C., IV (1988) Reproductive toxicity of tricresyl phosphate in a continuous breeding protocol in Swiss (CD-1) mice. *Fundam. appl. Toxicol.*, 10: 344-354.

CHAUDHURI, R.N. (1965) Paralytic disease caused by consumption of flour contaminated with tricresyl phosphate. *Trans. R. Soc. Trop. Med. Hyg.*, 59: 98-102.

CHAUDHURI, R.N., SEN GUPTA, P.C., CHAKRAVARTI, R.N., MUKHERJEE, A.M., GHOSH, S.M., CHATTERJEA, J.B., SARKAR, J.K., BASU, S.P., ADHYA, R.N., MITRA, N.K., SAHA, T.K., & MITRA, P. (1962) Recent outbreak of a paralytic disease in Malda, West Bengal. *Bull. Calcutta Sch. Trop. Med.*, 10: 141-152.

CHEMICAL AND GEOLOGICAL LABORATORIES LTD (1971) Calgary, Alberta, Chemical and Geological Laboratories (Laboratory report No. E71-5360).

CLAYTON, G.D. & CLAYTON, F.E., ed. (1981) *Patty's industrial hygiene and toxicology*, New York, Wiley-Interscience, Vol. 2A, pp. 2362-2363.

COHEN, S.D. & MURPHY, S.D. (1970) Comparative potentiation of malathion by triorthotolyl phosphate in four classes of vertebrates. *Toxicol. appl. Pharmacol.*, 16: 701-708.

DAFT, J.L. (1982) Identification of aryl/alkyl phosphate residues in foods. *Bull. environ. Contam. Toxicol.*, 29: 221-227.

DAGLEY, S. & PATEL, M.D. (1957) Oxidation of p-cresol and related compounds by a *Pseudomonas*. *Biochem. J.*, 66: 227-233.

DAVISON, A.N. (1953) The cholinesterases of the central nervous system after the administration of organo-phosphorus compounds. *Brit. J. Pharmacol.*, 8: 212-216.

DAWSON, G.W., JENNINGS, A.L., DROZDOWSKI, D., & RIDER, E. (1977) The acute toxicity of 47 industrial chemicals to fresh and salt water fishes. *J. hazard. Mater.*, 1: 303-318.

DEO, P.G. & HOWARD, P.H. (1978) Combined gas-liquid chromatographic mass spectrometric analysis of some commercial aryl phosphate oils. *J. Assoc. Off. Anal. Chem.*, 61: 266-271.

DRUYAN, E.A. (1975) [Separation and determination of tricresyl phosphate, triphenyl phosphate, phenol, o-, m-, and p-cresol by thin-layer chromatography.] *Gig. i Sanit.*, 10: 62-65 (in Russian).

DUKE, A.J. (1978) The significance of isomerism and complexity of composition on the performance of triaryl phosphate plasticisers in PVC. *Chimia*, 32: 457-463.

EAJ (1977) [*Environmental monitoring of chemicals*], Tokyo, Environment Agency Japan, pp. 212-214 (Environmental Survey Report Series, No. 3) (in Japanese).

EAJ (1979) [*Chemicals in the environment*], Tokyo, Environment Agency Japan, pp. 61, 78, 92, 93 (Office of Health Studies Report Series, No. 5) (in Japanese).

EAJ (1981) [*Environmental monitoring of chemicals: Environmental Survey Report of 1978, 1979 F.Y.*], Tokyo, Environment Agency Japan, pp. 116-117 (in Japanese).

ELDEFRAWI, A.T., MANSOUR, N.A., BRATTSTEN, L.B., AHRENS, V.D., & LISK, D.J. (1977) Further toxicologic studies with commercial and candidate flame retardant chemicals. Part II. *Bull. environ. Contam. Toxicol.*, 17: 720-726.

ETO, M., CASHIDA, J.E., & ETO, T. (1962) Hydroxylation and cyclization reactions involved in the metabolism of tri-o-cresyl phosphate. *Biochem. Pharmacol.*, 11: 337-352.

ETO, M., OSHIMA, Y., & CASHIDA, J.E. (1967) Plasma albumin as a catalyst in cyclization of diaryl-o-(a-hydroxy)tolyl phosphates. *Biochem. Pharmacol.*, 16: 295-308.

ETO, M., HASHIMOTO, Y., OZAKI, K., KASAI, T., & SASAKI Y. (1975) Fungitoxicity and insecticide synergism of monothioquinol phosphate esters and related compounds. *Botyu Kagaku*, 40: 110-117.

FINNEGAN, R.A. & MATSON, J.A. (1972) Irradiation of triaryl phosphate esters. A new photochemical coupling reaction. *J. Am. Chem. Soc.*, 94: 4780-4782.

FRANCHINI, I., CAVATORTA, A., D'ERRICO, M., DE SANTIS, M., ROMITA, G., GATTI, R., JUVARRA, G., & PALLA, G. (1978) Studies on the etiology of the experimental neuropathy from industrial adhesives (glues). *Experientia (Basel)*, 34: 250-252.

References

FREEMAN, G.B., IRWIN, R., TREJO, R., HEJTMANCIK, M., RYAN, M., & PETERS, A.C. (1988) Reversibility and tolerance to tricresyl phosphate-induced neurotoxic effects in F344 rats. *Toxicologist,* **8**: 76.

FUKUSHIMA, M. & KAWAI, S. (1986) [Present status and transition of selected organophosphoric acid triesters in the water area of Osaka city.] *Seitai Kagaku,* **8**: 13-24 (in Japanese).

GARTNER, W. & ELSAESSER, K.H. (1943) [Commercial ortho-tricresylphosphate poisoning.] *Arch. Gewerbepathol. Gewerbehyg.,* **12**: 1-9 (in German).

GEOFFROY, H., SLOMIC, A., BENEBADJI, M., & PASCAL, P. (1960) Myèlopolynéurites tri-crésyl phosphatées. Toxi-épidémie marocaine de septembre-octobre 1959. *World Neurol.,* **1**: 294-315.

GOLDFAIN, E. (1930) Jamaica ginger multiple neuritis. *J. Oklahoma State Med. Assoc.,* **23**: 191-193.

GOLDSTEIN, D.A., MCGUIGAN, M.A., & RIPLEY, B.D. (1988) Acute tricresyl phosphate intoxication in childhood. *Hum. Toxicol.,* **7**: 179-182.

GOODALE, R.H. & HUMPHREYS, M.B. (1931) Jamaica ginger paralysis. Autopsy observations. *J. Am. Med. Assoc.,* **96**: 14-16.

GROSS, E. & GROSSE, A. (1932) [A contribution to the toxicology of orthotricresylphosphates.] *Arch. exp. Pathol. Pharmakol.,* **168**: 473-514 (in German).

HAGGERTY, G.J., HEJTMANCIK, M., DESKIN, R., RYAN, M.J., PETERS, A.C., & IRWIN, R. (1986) Effects of subchronic exposure to tricresyl phosphate in F344 rats and B6C3F1 mice: behavioural and morphological correlates. *Toxicologist,* **6**: 218.

HATTORI, Y., ISHITANI, H., KUGE, Y., & NAKAMOTO, M. (1981) [Environmental fate of organic phosphate esters.] *Shuishitu Odaku Kenkyu,* **4**: 137-141 (in Japanese).

HAWORTH, S., LAWLOR, T., MORTELMANS, K., SPECK, N., & ZEIGER, E. (1983) Salmonella mutagenicity test results for 250 chemicals. *Environ. Mutagen., Suppl.* **1**: 3-142.

HINE, C.H., DUNLAP, M.K., RICE, E.G., COURSEY, M.M., GROSS, R.M., & ANDERSON, H.H. (1956) The neurotoxicity and anticholinesterase properties of some substituted phenyl phosphates. *J. Pharmacol. exp. Ther.,* **116**: 227-236.

HINE, C., ROWE, V.K., WHITE, E.R., DARMER, K.I., Jr, & YOUNGBLOOD, G.T. (1981) In: Clayton, G.D. & Clayton, F.E., ed. *Patty's industrial hygiene and toxicology,* 3rd revised ed., New York, Wiley-Interscience, Vol. 2A, pp. 2362-2363.

HJORTH, N. (1962) Contact dermatitis from cellulose acetate film. Cross-sensitization between tricresylphosphate (TCP) and triphenylphosphate (TPP). *Contact dermatitis,* **2**: 86-100.

HODGE, H.C. & STERNER, J.H. (1943) The skin absorption of triorthocresyl phosphate as shown by radioactive phosphorus. *J. Pharmacol. exp. Ther.*, 79: 225-234.

HOLLIFIELD, H.C. (1979) Rapid nephelometric estimate of water solubility of highly insoluble organic chemicals of environmental interest. *Bull. environ. Contam. Toxicol.*, 23: 579-586.

HOTSTON, R.D. (1946) Outbreak of polyneuritis due to orthotricresyl phosphate poisoning. *Lancet*, 1: 207.

HOWARD, P.H. & DEO, P.G. (1979) Degradation of aryl phosphates in aquatic environments. *Bull. environ. Contam. Toxicol.*, 22: 337-344.

HUDEC, T., THEAN, J., KUEHL, D., & DOUDHERTY, R.C. (1981) Tris(dichloropropyl)phosphate, a mutagenic flame retardant: Frequent occurrence in human seminal plasma. *Science*, 211: 951-952.

HUNTER, D., PERRY, K.M.A., & EVANS, R.B. (1944) Toxic polyneuritis arising during the manufacture of tricresyl phosphate. *Br. J. ind. Med.*, 1: 227-231.

INDEN, T. & TACHIBANA, S. (1975) [Damage of crops by gases from the plastic materials under covering conditions.] *Bull. Fac. Agric. Mie Univ.*, 1975: 1-10 (in Japanese).

INOUE, N., FUJISHIRO, K., MORI, K., & MATSUOKA, M. (1988) Triorthocresyl phosphate poisoning: A review of human cases. *Sangyo-Ika-Daigaku-Zasshi*, 10(4): 433-442.

IRWIN, R., FREEMAN, G.B., TREJO, R., HEJTMANCIK, M., & PETERS, A. (1987) Effects of chronic exposure to tricresyl phosphate in F344 rats and B6C3F$_1$ mice: behavioural and biochemical correlates. *Soc. Neurosci. Abstr.*, 13: 90.

ISHIKAWA, S., TAKETOMI, M., & SHINOHARA, R. (1985) Determination of trialkyl and triaryl phosphates in environmental samples. *Water Res.*, 19: 119-125.

IZMEROV, N.F., SANOTSKY, I.V., & SIDOROV, K.K. (1982) *Toxicometric parameters of industrial toxic chemicals under single exposure*, Moscow, Centre of International Projects, GKNT, p. 114.

JETER, H. (1930) Autopsy report of a case of so-called "Jake Paralysis". *J. Am. Med. Assoc.*, 95: 112-113.

JOHANSSEN, F.R., WRIGHT, P.L., GORDON, D.E., LEVINSKAS, G.J., RADUE, R.W., & GRAHAM, P.R. (1977) Evaluation of delayed neurotoxicity and dose-response relationships of phosphate esters in the adult hen. *Toxicol. appl. Pharmacol.*, 41: 291-304.

JOHNSON, M.K. (1969) Delayed neurotoxic action of some organophosphorus compounds. *Br. med. Bull.*, 25: 231-235.

References

JOHNSON, M.K. (1974) The primary biochemical lesion leading to the delayed neurotoxicity of some organophosphorus esters. *J. Neurochem.*, 23: 785-789.

JOHNSON, M.K. (1975a) Organophosphorus esters causing neurotoxic effects. Mechanism of action and structure/activity studies. *Arch. Toxicol.*, 34: 259-288.

JOHNSON, M.K. (1975b) The delayed neuropathy caused by some organophosphorus esters: mechanism and challenge. *CRC Crit. Rev. Toxicol.*, 3: 289-316.

JOHNSON, M.K. & BARNES, J.M. (1970) The sensitivity of chicks to the delayed neurotoxic effects of some organophosphorous compounds. *Biochem. Pharmacol.*, 19: 3045.

KANAZAWA, J. (1978) Studies on formulation and residue analysis of pesticides. *J. pestic. Sci.*, 3: 185-193.

KAWAI, S., FUKUSHIMA, M., ODA, K., & UNO, G. (1978) [Water pollution caused by organophosphorus compounds.] *Kankyo Gijyutsu*, 7: 668-675 (in Japanese).

KENMOTSU, K., SAITO, N., OGINO, N., & MATSUNAGA, K. (1979) [An environmental survey of chemicals. XI. Analytical methods of organic phosphoric acid triesters.] *Okayama-ken Kankyo Hoken Senta Nempo*, 3: 175-191 (in Japanese).

KENMOTSU, K., MATSUNAGA, K., & ISHIDA, T. (1980a) [Multiresidue determination of phosphoric acid triesters in fish, sea sediment and sea water.] *J. Food Hyg. Soc. Jpn.*, 21: 18-31 (in Japanese).

KENMOTSU, K., MATSUNAGA, K., & ISHIDA, T. (1980b) [Studies on the mechanisms of biological activities of various environmental pollutants. V. Environmental fate of organic phosphoric acid triesters.] *Okayama-ken Kankyo Hoken Senta Nempo*, 4: 103-110 (in Japanese).

KENMOTSU, K., MATSUNAGA, K., & ISHIDA, T. (1981a) [Studies on the biological toxicity of several pollutants in environments.] *Okayama-ken Kankyo Hoken Senta Nempo*, 5: 167-175 (in Japanese).

KENMOTSU, K., MATSUNAGA, K., SAITO, N., & OGINO, Y. (1981b) [An environmental survey of chemicals. XVII. Multiresidue determination of organic phosphate esters in environment samples.] *Okayama-ken Kankyo Hoken Senta Nempo*, 5: 145-156 (in Japanese).

KENMOTSU, K., MATSUNAGA, K., SAITO, N., OGINO, Y., & ISHIDA, T. (1982a) [An environmental survey of chemicals. XIX. Determination of organophosphoric acid triesters (2).] *Okayama-ken Kankyo Hoken Senta Nempo*, 6: 126-132 (in Japanese).

KENMOTSU, K., MATSUNAGA, K., SAITO, N., OGINO, Y., & ISHIDA, T. (1982b) [Studies on the biological toxicity of several pollutants in environments. VII. GC/MS Spectrometric determination of organophosphoric acid triesters in sediment.] *Okayama-ken Kankyo Hoken Senta Nempo*, 6: 142-152 (in Japanese).

KENMOTSU, K., NAKAGIRI, M., OGINO, Y., MATSUNAGA, K., & ISHIDA, T. (1983) [An environmental survey of chemicals. XXII. GC/MS spectrometric determination of organophosphoric acid triesters in sediment (2).] *Okayama-ken Kankyo Hoken Senta Nempo*, 7: 143-149 (in Japanese).

KIMMERLE, G. & LOESER, E. (1974) Delayed neurotoxicity of organophosphorus compounds and copper concentration in the serum of hens. *Environ. Qual. Saf.*, 3: 174-178.

KONASEWICH, D., TRAVERSY, W., & ZAR, H. (1978) Status report on organic and heavy metal contaminants in the Lakes Erie, Michigan, Huron and Superior basins. *Great Lakes Water Qual. Bd.*

KRISHNAMURTHY, M.N., RAJALAKSHMI, S., & PRAKASH KAPUR, O. (1985) Detection of tricresyl phosphates and determination of tri-o-cresyl phosphate in edible oils. *J. Assoc. Off. Anal. Chem.*, **68**(6): 1074-1076.

KU, Y. & ALVAREZ, G.H. (1982) Biodegradation of (^{14}C)tri-p-cresyl phosphate in a laboratory activated-sludge system. *Appl. environ. Microbiol.*, **43**: 619-622.

KUREBAYASHI, H., TANAKA, A., & YAMAHA, T. (1985) Metabolism and disposition of the flame retardant plasticizer, tri-p-cresyl phosphate, in the rat. *Toxicol. appl. Pharmacol.*, **77**: 395-404.

LAPP, T.W. (1976) *Study on chemical substances from information concerning the manufacture, distribution, use, disposal, alternatives, and magnitude of exposure to the environment and man: Task 1 - The manufacture and use of selected aryl and alkyl phosphate esters*, Springfield, Virginia, National Technical Information Service, 129 pp (EPA contract No. 68-01-2687) (NTIS PB-251819).

LEBEL, G.L. & WILLIAMS, D.T. (1983) Determination of organic phosphate triesters in human adipose tissue. *J. Assoc. Off. Anal. Chem.*, **66**: 691-699.

LEBEL, G.L., WILLIAMS, D.T., GRIFFITH, G., & BENOIT, F.M. (1979) Isolation and concentration of organophosphorus pesticides from drinking water at the ng/L level, using macroreticular resin. *J. Assoc. Off. Anal. Chem.*, **62**: 241-249.

LEBEL, G.L., WILLIAMS, D.T., & BENOIT, F.M. (1981) Gas chromatographic determination of trialkyl/aryl phosphates in drinking water, following isolation using macroreticular resin. *J. Assoc. Off. Anal. Chem.*, **64**: 991-998.

LEFAUX, R. (1968) *Practical toxicology of plastics*, London, Iliffe Books Ltd, p. 580.

LEFAUX, R. (1972) Industrial toxicology. Phosphoric esters. In: *Practical toxicology of plastics*, London, Iliffe Books Ltd, pp. 121-127.

LEVEQUE, J. (1983) Tricresyl phosphates. In: *Encyclopaedia of occupational health safety*, 3rd ed., Geneva, International Labour Office, Vol. 2, pp. 2216-2218.

LOCKHART, W.L., WAGEMANN, R., CLAYTON, J.W., GRAHAM, B., & MURRAY, D. (1975) Chronic toxicity of a synthetic tri-aryl phosphate oil to fish. *Environ. Physiol. Biochem.*, **5**: 361-369.

LOMBARDO, P. & EGRY, I.J. (1979) Identification and gas-liquid chromatographic determination of aryl phosphate residues in environmental samples. *J. Assoc. Off. Anal. Chem.*, **62**: 47-51.

LOROT, C. (1899) Les combinations de la créosote dans le traitement de la tuberculose pulmonaire, Paris (Thèse de Paris No. 25).

LU, P.-Y. & METCALF, R.L. (1975) Environmental fate and biodegradability of benzene derivatives as studied in a model aquatic ecosystem. *Environ. Health Perspect.*, **10**: 269-284.

MAJORS, R.E. & JOHNSON, E.L. (1978) High-performance exclusion chromatography of low-molecular-weight additives. *J. Chromatogr.*, **167**: 17-30.

MARHOLD, J.V. (1972) *Collection of results of toxicological assessments of compounds and formulations tested at the Department of Toxicology, Research Institute for Organic Syntheses,* Prague, Institute for Management Training in the Chemical Industry.

MAUSNER, M., BENEDICT, J.H., BOOMAN, K.A., BRENNER, T.E., CONWAY, R.A., DUTHIE, J.R., GARRISON, L.J., HENDRIX, C.D., & SHEWMAKER, J.E. (1969) The status of biodegradability testing of nonionic surfactants. *J. Am. Oil Chem. Soc.*, **46**: 432-440.

MAYER, F.L. & ELLERSIECK, M.R. (1986) *Manual of acute toxicity: Interpretation and data base for 410 chemicals and 66 species of freshwater animals,* Springfield, Virginia, National Technical Information Service.

MAYER, F.L., ADAMS, W.J., FINLEY, M.T., MICHAEL, P.R., MEHRLE, P.M., & SAEGER, V.W. (1981) Phosphate ester hydraulic fluids: An aquatic environmental assessment of Pydrauls 50E and 115E. In: Branson R. & Dickson, K.L., ed. *Aquatic Toxicology and Hazard Assessment: Fourth Conference,* Philadelphia, American Society for Testing and Materials, pp. 103-123 (ASTM STP 737).

MELE, J.M. & JENSH, R.P. (1977) Teratogenic effects of orally administered tri-o-cresyl phosphate on Wistar albino rats. *Teratology*, **15**: 32A.

MERRITT, H.H. & MOOR, M. (1930) Peripheral neuritis associated with ginger extract ingestion. *New Engl. J. Med.*, **203**: 4-12.

MERTENS, H.G. (1948) [Clinical aspects of tri-orthocrésyl phosphate poisoning (including a report on several cases of poisoning in an Igelit factory).] *Arch. Psychiatr. Nervenkr.*, **179**: 458-482 (in German).

MRI (1979) *Assessment of the need for limitation on triaryl and trialkyl/aryl phosphates. Draft final report,* Kansas City, Midwest Research Institute (EPA Contract 68-01-4313).

MODERN PLASTICS ENCYCLOPEDIA (1975) International Advertising Supplement 52(10A), p 697, New York, McGraw-Hill Inc.

MORAZAIN, R. & ROSENBERG, P. (1970) Lipid changes in tri-o-cresyl-phosphate-induced neuropathy. *Toxicol. appl. Pharmacol.*, 16: 461-474.

MORGAN, A.A. & HUGHES, J.P.W. (1981) An investigation into the value of cholinesterase estimations of workers in a plant manufacturing tri-aryl phosphate plasticizers. *J. Soc. Occup. Med.*, 31: 69-75.

MORGAN, J.P. (1982) The Jamaica Ginger Paralysis. *J. Am. Med. Assoc.*, 248: 1864-1867.

MORGAN, J.P. & PENOVICH, P. (1978) Jamaica Ginger Paralysis. Forty-seven-year follow-up. *Arch. Neurol.*, 35: 530-532.

MORRISSEY, R.E., SCHWETZ, B.A., LAMB, J.C., IV, ROSS, M.D., TEAGUE, J.L., & MORRIS, R.W. (1988a) Evaluation of rodent sperm, vaginal cytology, and reproductive organ weight data from National Toxicology Program 13-week study. *Fundam. appl. Toxicol.*, 11: 343-358.

MORRISSEY, R.E., SCHWETZ, B.A., LAMB, J.C., IV, ROSS, M.D., TEAGUE, J.L., & MORRIS, R.W. (1988b) Association of sperm, vaginal cytology, and reproductive organ weight data with results of continuous breeding reproductive studies in Swiss CD-1 mice. *Fundam. appl. Toxicol.*, 11: 359-371.

MUIR, D.C.G. (1984) Phosphate esters. In: Hutzinger, O., ed. *The handbook of environmental chemistry*, Berlin, Heidelberg, New York, Tokyo, Springer-Verlag, Vol. 3, Part C, pp. 41-66.

MUIR, D.C.G. & GRIFT, N.P. (1983) Extraction and cleanup procedures for determination of diarylphosphates in fish, sediment, and water samples. *J. Assoc. Anal. Off. Chem.*, 66: 684-690.

MUIR, D.C.G., GRIFT, N.P., & SOLOMON, J. (1980) Determination of several triarylphosphates in fish and sediment samples. *Can. Plains Proc.*, 9: 1-12.

MUIR, D.C.G., GRIFT, N.P., & SOLOMON, J. (1981) Extraction and cleanup of fish, sediment, and water for determination of triaryl phosphates by gas-liquid chromatography. *J. Assoc. Off. Anal. Chem.*, 64: 79-84.

MUIR, D.C.G., GRIFT, N.P., & LOCKHART, W.L. (1982) Comparison of laboratory and field results for prediction of the environmental behavior of phosphate esters. *Environ. Technol. Chem.*, 1: 113-119.

MUIR, D.C.G., YARECHEWSKI, A.L., & GRIFT, N.P. (1983) Environmental dynamics of phosphate esters. III. Comparison of the bioconcentration of four triaryl phosphates by fish. *Chemosphere*, 12: 155-166.

MURRAY, D.A.J. (1975) Analysis of tri-aryl phosphate esters and the determination of IMOL S-140 in fish tissue and water samples by gas chromatography. *J. Fish Res. Board Can.*, 32: 457-460.

MYERS, D.K., REBEL, J.B.J., VEEGER, C., KEMP, A., & SIMONS, E.G.L. (1955) Metabolism of triaryl phosphates in rodents. *Nature (Lond.)*, 176: 259-260.

NEELY, W.B., BRANSON, D.R., & BLAU, G.E. (1974) Partition coefficient to measure the bioconcentration potential of organic chemicals in fish. *Environ. Sci. Technol.*, **8**: 1113-1115.

NEVINS, M.J. & JOHNSON, W.W. (1978) Acute toxicity of phosphate ester mixtures to invertebrates and fish. *Bull. environ. Contam. Toxicol.*, **19**: 250-256.

NICHOLSON, S.S. (1974) Bovine posterior paralysis due to organophosphate poisoning. *J. Am. Vet. Med. Assoc.*, **165**: 280-281.

NOBILE, E.R., PAGE, S.W., & LOMBARDO, P. (1980) Characterization of four commercial flame retardant aryl phosphates. *Bull. environ. Contam. Toxicol.*, **25**: 755-761.

NOMEIR, A.A. & ABOU-DONIA, M.B. (1983) High-performance liquid chromatographic analysis on radial compression column of the neurotoxic tri-o-cresyl phosphate and metabolites. *Anal. Biochem.*, **135**: 296-303.

NOMEIR, A.A. & ABOU-DONIA, M.B. (1984) Disposition of (^{14}C)tri-o-cresyl phosphate and its metabolites in various tissues of the male cat following a single dermal application. *Drug Metab. Disp.*, **12(6)**: 705-711.

NOMEIR, A.A. & ABOU-DONIA, M.B. (1986a) Studies on the metabolism of the neurotoxic tri-o-cresyl phosphate synthesis and identification by infrared, proton nuclear magnetic resonance and mass spectrometry of five of its metabolites. *Toxicology*, **38**: 1-13.

NOMEIR, A.A. & ABOU-DONIA, M.B. (1986b) Studies on the metabolism of the neurotoxic tri-*o*-cresyl phosphate, distribution, excretion, and metabolism in male cats after a single, dermal application. *Toxicology*, **38**: 15-33.

OECD (1981) *OECD guidelines for testing of chemicals. Section 2: Effects on biotic systems*, Paris, Organization for Economic Cooperation and Development, Publications Office.

OFSTAD, E.B. & SLETTEN, T. (1985) Composition and water solubility determination of a commercial tricresylphosphate. *Sci. total Environ.*, **43**: 233-241.

OISHI, H., OISHI, S., & HIRAGA, K. (1982) Toxicity of several phosphoric acid esters in rats. *Toxicol. Lett.*, **13**: 29-34.

OLSON, B.A. & BURSIAN, S.J. (1988) Effect of route of administration on the development of organophosphate-induced delayed neurotoxicity in 4-week-old chicks. *J. Toxicol. environ. Health*, **23(4)**: 499-505.

PACIOREK, K.J.L., KRATZER, R.H., KAUFMAN, J., NAKAHARA, J.H., CHRISTOS, T., & HARTSTEIN, A.M. (1978) Thermal oxidative degradation studies of phosphate esters. *Am. Ind. Hyg. Assoc. J.*, **39**: 633-639.

PADILLA, S. & VERONESI, B. (1985) The relationship between neurological damage and neurotoxic esterase inhibition in rats acutely exposed to tri-ortho-cresyl phosphate. *Toxicol. appl. Pharmacol.*, **78(1)**: 78-87.

PARNITZKE, K.H. (1946) [On the recent accumulation of ortho-tricresylphosphate poisoning (sources and clinical course).] *Dtsch. Gesundheitswes.*, **1**: 666-670 (in German).

PEGUM, J.S. (1966) Contact dermatitis from plastics containing tri-aryl phosphates. *Br. J. Dermatol.*, **78**: 626-631.

PENMAN, D.R & OSBORNE, G.O. (1976) Trialkyl phosphates and related compounds as antifertility agents of the twospotted spider mite. *J. econ. Entomol.*, **69**(2): 266-268.

PICKARD, M.A., WHELIHAN, J.A., & WESTLAKE, D.W.S. (1975) Utilization of triaryl phosphates by a mixed bacterial population. *Can. J. Microbiol.*, **21**: 140-145.

PLAPP, F.W., Jr & TONG, H.H.C. (1966) Synergism of malathion and parathion against resistant insects: Phosphorus esters with synergistic properties. *J. econ. Entomol.*, **59**(1): 11-15.

PRENTICE, D.E. & MAJEED, S.K. (1983) A subchronic study (90 days) using multiple dose levels of tri-ortho-cresyl phosphate (TOCP): some neuropathological observations in the domestic hen. *Neurotoxicology*, **4**(2): 277-282.

PRINEAS, J.S. (1969) The pathogenesis of dying-back polyneuropathies. Part I. An ultrastructural study of experimental tri-ortho-cresyl phosphate intoxication in the cat. *J. Neuropathol. exp. Neurol.*, **28**: 571-597.

QUINSTAD, G.B., STAIGER, L.E., & SCHOOLEY, D.A. (1975) Environmental degradation of the insect growth regulator methoprene. V. Metabolism by houseflies and mosquitoes. *Pestic. Biochem. Physiol.*, **5**: 233-241.

RAMSEY, J.D. & LEE, T.D. (1980) Gas-liquid chromatographic retention indices of 296 non-drug substances on SE-30 or OV-1 likely to be encountered in toxicological analyses. *J. Chromatogr.*, **184**: 185-206.

RENBERG, L., SUNDSTROM, G., & SUNDH-NYGARD, K. (1980) Partition coefficients of organic chemicals derived from reversed phase thin layer chromatography. Evaluation of methods and application on phosphate esters, polychlorinated paraffins and some PCB-substitutes. *Chemosphere*, **9**: 683-691.

ROBERTS, N.L., FAIRLEY, C., & PHILLIPS, C. (1983) Measurements and evaluation of clinical signs following use of the hen as the animal model for screening acute delayed and subchronic neurotoxicity studies. *Neurotoxicology*, **4**(2): 263-270.

ROBERTSON, D.G., SCHWAB, B.W., SILLS, R.D., RICHARDSON, R.J., & ANDERSON, R.J. (1987) Electrophysiologic changes following treatment with organophosphorus-induced delayed neuropathy-producing agents in the adult hen. *Toxicol. appl. Pharmacol.*, **87**: 420-429.

ROGER, H. & RECORDIER, M. (1934) Les polynéurites phosphocréosotiques (phosphate de créosote, ginger paralysis, apiol). *Ann. Med.*, **35**: 44-63.

SAEGER, V.W., HICKS, O., KALEY, R.G., MICHAEL, P.R., MIEURE, J.P., & TUCKER, E.S. (1979) Environmental fate of selected phosphate esters. *Environ. Sci. Technol.*, 13: 840-844.

SAITO, C., KATO, T., TANIGUCHI, H., FUJITA, T., WADA, H., & MORI, Y. (1974) [Subacute toxicity of tricresylphosphate (TCP) in rats.] *Pharmacometrics*, 8: 107-118 (in Japanese).

SAMPSON, B.F. (1942) The strange Durban epidemic of 1937. *S. Afr. med. J.*, 16: 1-9.

SASAKI, K., SUZUKI, T., TAKEDA, M., & UCHIYAMA, M. (1982) Bioconcentration and excretion of phosphoric acid triesters by Killifish *(Oryzeas latipes)*. *Bull. environ. Contam. Toxicol.*, 28: 752-759.

SCHWAB, H. (1948) [Psychic disturbance by triorthocresylphosphate poisoning.] *Dtsch. Med. Wochenschr.*, 73: 124-125 (in German).

SENANAYAKE, N. (1981) Tri-cresyl phosphate neuropathy in Sri Lanka: a clinical and neurophysiological study with a three year follow up. *J. Neurol. Neurosurg. Psychiatry*, 44: 775-780.

SENANAYAKE, N. & JEYARATNAM, J. (1981) Toxic polyneuropathy due to gingili oil contaminated with tri-cresyl phosphate affecting adolescent girls in Sri Lanka. *Lancet, 10 January:* 88-89.

SHARMA, R.P. & WATANABE, P.G. (1974) Time related disposition of tri-o-tolyl phosphate (TOTP) and metabolites in chicken. *Pharmacol. Res. Commun.*, 6(5): 475-484.

SHELDON, L.S. & HITES, R.A. (1978) Organic compounds in the Delaware River. *Environ. Sci. Technol.*, 12: 1188-1194.

SHELDON, L.S. & HITES, R.A. (1979) Sources and movement of organic chemicals in the Delaware River. *Environ. Sci. Technol.*, 13: 574-579.

SITTHICHAIKASEM, S. (1978) *Some toxicological effects of phosphate esters on rainbow trout and bluegill*, Iowa State University (Ph. D. Thesis).

SMITH, H.V. & SPALDING, J.M.K. (1959) Outbreak of paralysis in Morocco due to ortho-cresyl phosphate poisoning. *Lancet*, 2: 1019-1021.

SMITH, M.I. & LILLIE, R.D. (1931) The histopathology of triorthocresyl phosphate poisoning. *Arch. Neurol Psychiatry*, 26: 976-992.

SMITH, M.I., ELVOVE, E., & FRAZIER, W.H. (1930) The pharmacological action of certain phenol esters, with special reference to the etiolgy of so-called ginger paralysis. *Public Health Rep.*, 45: 2509-2524.

SMITH, M.I., ENGEL, E.W., & STOHLMAN, F.F. (1932) Further studies on the pharmacology of certain phenol esters with special reference to the relation of chemical constitution and physiologic action. *Natl. Inst. Health Bull.*, 160: 1-53.

SOMKUTI, S.C., LAPADULA, D.M., CHAPIN, R.E., LAMB, J.C., IV, & ABOU-DONIA, M.B. (1987a) Reproductive tract lesions resulting from subchronic administration (63 days) of tri-o-cresyl phosphate in male rats. *Toxicol. appl. Pharmacol.*, **89**: 49-63.

SOMKUTI, S.C., LAPADULA, D.M., CHAPIN, R.E., LAMB, J.C., IV, & ABOU-DONIA, M.B. (1987b) Time course of the tri-o-cresyl phosphate-induced testicular lesion in F-344 rats: enzymatic, hormonal, and sperm parameter studies. *Toxicol. appl. Pharmacol.*, **89**: 64-72.

SOMKUTI, S.G., LAPADULA, D.M., CHAPIN, R.E., & ABOU-DONIA, M.B. (1987c) Testicular toxicity following oral administration of tri-o-cresyl phosphate (TOCP) in roosters. *Toxicol. Lett.*, **37**: 279-290.

SOMKUTI, S.G., TILSON, H.A., BROWN, H.R., CAMPBELL, G.A., LAPADULA, D.M., & ABOU-DONIA, M.B. (1988) Lack of delayed neurotoxic effect after tri-ortho-cresyl phosphate treatment in male Fischer-344 rats: biochemical, neurobehavioural, and neuropathological studies. *Fundam. appl. Toxicol.*, **10**: 199-205.

SOROKIN, M. (1969) Orthocresyl phosphate neuropathy: Report of an outbreak in Fiji. *Med. J. Aust.*, **1**: 506-508.

SPOERRI, P.E. & GLEES, P. (1979) Ultrastructural reactions of spinal ganglia to tri-ortho-cresyl phosphate: Effects of neurotoxicity. *Cell Tissue Res.*, **199**: 409-414.

SPOERRI, P.E. & GLEES, P. (1980) Ultrastructural changes in neurons and neuroglia of the avian telencephalon (Hyperstriatum Accessorium) following tri-ortho-cresyl phosphate intoxication. *Cell Tissue Res.*, **206**: 203-210.

STAEHELIN, R. (1941) [On the triorthocresylphosphate poisonings.] *Schweiz. Med. Wochenschr.*, **71**: 1-5 (in German).

STUMPF, A.M., TANAKA, D.JR., AULERIOH, A.J., & BURSIAN, S.J. (1989) Delayed neurotoxic effects of tri-o-tolyl phosphate in the European ferret. *J. Toxicol. environ. Health*, **26**: 61-73.

STURM, R.N. (1973) Biodegradability of nonionic surfactants: Screening test for predicting rate and ultimate biodegradation. *J. Am. Oil Chem. Soc.*, **50**: 159-167.

SUGDEN, E.A., GREENHALGH, R., & PETTIT, J.R. (1980) Characterization of delayed neurotoxic triaryl phosphates by analysis of trifluoroacetylated phenolic moieties. *Environ. Sci. Technol.*, **14**(12): 1498-1501.

SUSSER, M. & STEIN, Z. (1957) An outbreak of tri-ortho-cresyl phosphate (TOCP) poisoning in Durban. *Br. J. ind. Med.*, **14**: 111-120.

SVENNILSON, E. (1960) Studies of triorthocresyl phosphate neuropathy, Morocco 1960. *Acta psychiatr. Scand.*, **150**(Suppl): 334-336.

TABERSHAW, I.R., KLEINFELD, M., & FEINER, B. (1957) Manufacture of tricresyl phosphate and other alkyl phenyl phosphates: An industrial hygiene study. *Am. Med. Assoc. Arch. ind. Health*, 15: 537-540.

TAYLOR, J.D. & BUTTAR, H.S. (1967) Evidence for the presence of 2-(o-cresyl)-4H-1:3:2-benzodioxaphosphoran-2-one in cat intestine following tri-o-cresyl phosphate administration. *Toxicol. appl. Pharmacol.*, 11: 529-537.

TER BRAAK, J.W.G. & CARRILLO, R. (1932) [Polyneuritis after usage of an abortifacient (tri-ortho-cresyl-phosphate-poisoning).] *Dtsch. Z. Nervenheilkd.*, 125: 86-116 (in German).

THOMPSON, J.E. & DUTHIE, J.R. (1968) The biodegradability and treatability of NTA. *J. Water Pollut. Control Fed.*, 40: 306-319.

TITTARELLI, P. & MASCHERPA, A. (1981) Liquid chromatography with graphite furnace atomic absorption spectrophotometric detector for speciation of organophosphorus compounds. *Anal. Chem.*, 53: 1466-1469.

TOCCO, D.R., RANDALL, J.L., YORK, R.G., & SMITH, M.K. (1987) Evaluation of the teratogenic effects of tri-ortho-cresyl phosphate in the Long-Evans hooded rat. *Fundam. appl. Toxicol.*, 8(3): 291-297.

US NIOSH (1977) *NIOSH manual of analytical methods*, 2nd ed., Cincinnati, Ohio, National Institute for Occupational Safety and Health, Vol. 3, P & CAM, pp. S/209-S/210 (DHEW (NIOSH) Publication No. 7-157C).

US NIOSH (1979) *Industrial hygiene walk-through survey report on organophosphorus exposures at Chevron Chemical, Belle Chase, Louisiana*, Cincinnati, Ohio, National Institute for Occupational Safety and Health (NTIS PB-82-216268.

US NIOSH (1980) *Industrial hygiene walk-through survey report on organophosphorus exposures at Rochester products division, General Motors Corp., 1000 Lexington Avenue, Rochester, New York*, Cincinnati, Ohio, National Institute for Occupational Safety and Health (PB82-104530, NTIS).

US NIOSH (1982) *Industrial hygiene walk-through survey report on organophosphorous exposures at Chevron Chemical, Belle Chase, Louisiana*, Cincinnati, Ohio, National Institute for Occupational Safety and Health, Division of Surveillance, Hazard Evaluations and Field Studies (Report No. IWA-89.10).

US SOAP AND DETERGENT ASSOCIATION (1965) A procedure and standards for the determination of the biodegradability of alkyl benzene sulfonate and linear alkylate sulfonate (Report of the Subcommittee on Biodegradation Test Methods). *J. Am. Oil Chem. Soc.*, 42: 986-993.

VASILESCU, C. & FLORESCU, A. (1980) Clinical and electrophysiological study of neuropathy after organophosphorus compound poisoning. *Arch. Toxicol.*, 43: 305-315.

VASWANI, M., MAHAJAN, P.M., SETIA, R.C., & BHIDE, N.K. (1983) A simple colorimetric method for estimation of tricresyl phosphate in edible oils. *J. Oil Technol. Assoc. India*, 15(1): 12-13.

VEITH, G.D., DEFOE, D.L., & BERGSTEDT, B.V. (1979) Measuring and estimating the bioconcentration factor of chemicals in fish. *J. Fish Res. Board Can.*, 36: 1040-1048.

VERONESI, B. (1984) A rodent-model of organophosphorus-induced delayed neuropathy: distribution of central (spinal cord) and peripheral nerve damage. *Neuropathol. appl. Neurobiol.*, 10(6): 357-368.

VERONESI, B. & ABOU-DONIA, M.B. (1982) Central and peripheral neuropathology induced in rats by tri-o-cresyl phosphate (TOCP). *Vet. hum. Toxicol.*, 24: 3 (abstract).

VERONESI, B., NEWLAND, D., & INMAN, A. (1984) The effect of metabolic interference on rodent-sensitivity to tri-ortho-cresyl phosphate. *Toxicologist*, 4: 55.

VICK, R.D., JUNK, G.A., AVERY, M.J., RICHARD, J.J., & SVEC, H.J. (1978) Organic emissions from combustion of combination coal/refuse to produce electricity. *Chemosphere*, 7: 893-902.

VONDERAHE, A.R. (1931) Pathologic changes in paralysis caused by drinking Jamaica Ginger. *Arch. Neurol. Psychiatry*, 25: 29-43.

VORA, D.D., DASTUR, D.K., BRAGANCA, B.M., PARIHAR, L.M., IYER, C.G.S., FONDEKAR, R.B., & PRABHAKARAN, K. (1962) Toxic polyneuritis in Bombay due to ortho-cresyl-phosphate poisoning. *J. Neurol. Neurosurg. Psychiatry*, 25: 234-242.

WAGEMANN, R. (1975) Some environmental and toxicological aspects of a tri-aryl phosphate synthetic oil. *Verh. Int. Ver. Limnol.*, 19: 2178-2184.

WAGEMANN, R., GRAHAM, R., & LOCKHART, W.L. (1974) *Studies on chemical degradation and fish toxicity of a synthetic tri-aryl phosphate lubrication oil, IMOL S-140*. Ottawa, Environment Canada, Fisheries and Marine Service, (Technical Report, No. 486).

WAKABAYASHI, A. (1980) [Environmental pollution caused by organophosphoric fire-proofing plasticizers.] *Annu. Rep. Tokyo Metrop. Res. Inst. Environ. Prot.*, 11: 110-113 (in Japanese).

WALTHARD, K.M. (1945) Aperçu des résultats obtenus lors des derniers examens des malades intoxiqués en 1940 par le phosphate tri-orthocrésilique. *Schweiz. Arch. Neurol. Psychiatry*, 58: 189-194.

WIGHTMAN, R.H. & MALAIYANDI, M. (1983) Physical properties of some synthetic trialkyl/aryl phosphate commonly found in environmental samples. *Environ. Sci. Technol.*, 17: 256-261.

WILLIAMS, D.T. & LEBEL, G.L. (1981) A national survey of tri(haloalkyl)-, trialkyl-, and triarylphosphates in Canadian drinking water. *Bull. environ. Contam. Toxicol.*, 27: 450-457.

References

WILLIAMS, D.T., NESTMANN, E.R., LEBEL, G.L., BENOIT, F.M., & OTSON, R. (1982) Determination of mutagenic potential and organic contaminants of Great Lakes drinking water. *Chemosphere*, 11: 263-276.

WILSON, R.D., ROWE, L.D., LOVERING, S.L., & WITZEL, D.A. (1982) Acute toxicity of tri-ortho-cresyl phosphate in sheep and swine. *Am. J. vet. Res.*, 43(11): 1954-1957.

WINDHOLZ, M., ed. (1983) *The Merck index*, 10th ed., Rahway, New Jersey, Merck and Co., Inc.

WOLFE, N.L. (1980) Organophosphate and organophosphorothionate esters: Application of linear free energy relationships to estimate hydrolysis rate constants for use in environmental fate assessment. *Chemosphere*, 9: 571-579.

WONG, P.T.S. & CHAU, Y.K. (1984) Structure-toxicity of triaryl phosphates in freshwater algae. *Sci. total Environ.*, 32(2): 157-165.

YASUDA, H. (1980) [Concentration of organic phosphorus pesticides in the atmosphere above the Dogo plain and Ozu basin.] *J. Chem. Soc. Jpn*, 1980(4): 645-653 (in Japanese).

ZELIGS, M.A. (1938) Upper motor neuron sequelae in "Jake" paralysis : A clinical follow up study. *J. nerv. ment. Dis.*, 87: 464-470.

ZITKO, V. (1980) Proceedings of the 6th Annual Aquatic Toxicity Workshop. *Can. Tech. Rep. fish aquat. Sci.*, 575: 234-265.

RESUME

1. Identité, propriétés physiques et chimiques, méthodes d'analyse

Le phosphate de tricrésyle (TCP) est un liquide visqueux, ininflammable, inexplosible et incolore. Son coefficient de partage entre l'octanol et l'eau (log de p_{ow}) est égal à 5,1. Il s'hydrolyse facilement en milieu alcalin pour donner du phosphate de dicrésyle, mais il est stable en milieu neutre ou acide à température normale.

Du point de vue analytique, la méthode de choix est la chromatographie en phase gazeuse avec détection par un dispositif sensible à l'azote/phosphore ou par photométrie de flamme. La limite de détection dans les échantillons aqueux est d'environ 1 mg/litre. Le TCP s'extrait facilement des solutions aqueuses au moyen de divers solvants organiques. Pour la purification, on emploie habituellement une colonne chromatographique de Florisil, mais il est difficile de séparer le TCP des lipides par cette méthode. D'autres méthodes de purification ont été recommandées à cette fin (chromatographie en phase gazeuse, chromatographie sur charbon activé ou Sep-pack C-18). Les réactifs analytiques sont souvent contaminés par des traces de TCP en raison de la très large utilisation faite de ce produit. Aussi faut-il prendre certaines précautions si l'on veut que l'analyse des traces de TCP soit fiable.

2. Sources d'exposition humaine et environnementale

Le TCP est généralement produit par réaction des crésols sur l'oxychlorure de phosphore. Il y a deux sources de production industrielle de crésols: l'"acide crésylique", résidu des fours à coke et du raffinage du pétrole, et les "crésols de synthèse" préparés par oxydation et dégradation du cymène. Le TCP est donc un mélange de divers phosphates de triaryle.

Le TCP est utilisé comme plastifiant des matières plastiques vinyliques, comme retardateur d'inflammation, comme additif pour les lubrifiants à très haute pression

Résumé

et comme liquide ininflammable dans les systèmes hydrauliques.

3. Transport, distribution et transformation dans l'environnement

La libération de TCP dans l'environnement est peu importante au cours de la production et se produit essentiellement lors de l'utilisation finale du produit. On estime qu'en 1977, 32 800 tonnes en ont été libérées au total dans l'atmosphère aux Etats-Unis.

En raison de sa faible solubilité dans l'eau et de sa forte adsorption aux particules, le TCP s'adsorbe rapidement aux sédiments des rivières et des lacs et aux particules du sol. Il est rapidement biodégradé en milieu aquatique, sa décomposition étant pratiquement complète dans les cours d'eau en l'espace de cinq jours. L'isomère ortho se décompose légèrement plus vite que les isomères méta et para. Le TCP est facilement biodégradé dans les boues d'égouts, avec une demi-vie de 7,5 heures, la dégradation atteignant 99% en l'espace de 24 heures. La dégradation abiotique est plus lente, puisque dans ce cas la demi-vie est de 96 jours.

Des facteurs de bioconcentration de 165-2768 ont été mesurés en laboratoire sur plusieurs espèces de poissons à l'aide de TCP radio-marqué. La radioactivité a rapidement disparu après cessation de l'exposition, la demi-vie de dépuration allant de 25,8 à 90 heures.

4. Niveaux dans l'environnement et exposition humaine

On a relevé dans l'air des concentration de TCP allant jusqu'à 70 ng/m^3 au Japon, les concentrations maximales dans une unité de production des Etats-Unis d'Amérique n'étant que de 2 ng/m^3. Dans un atelier de remplissage de fûts d'huile lubrifiante, aux Etats-Unis d'Amérique, on a constaté que l'air ne contenait que 0,8 mg/m^3 de TCP, la concentration ne dépassant pas 0,15 mg/m^3 en phosphates totaux, dans une unité de moulage de zinc d'une usine d'automobiles. Les concentrations mesurées dans l'eau de boisson au Canada se sont révélées faibles (0,4 à 4,3 ng/litre) et le TCP n'a pas pu être décelé dans l'eau des puits. Dans les rivières et les lacs, les concentrations sont souvent beaucoup plus élevées. Toutefois on peut attribuer cet état de choses à la présence de sédiments en suspension auxquels le TCP est fortement adsorbé.

Les concentrations sédimentaires sont plus élevées puisqu'elles peuvent atteindre 1300 ng/g dans le sédiments de cours d'eau et 2160 ng/g par les sédiments marins.

Des concentrations élevées ont été mesurées dans le sol et la végétation aux alentours d'unités de production.

On a fait état de résidus dans des poissons et des fruits de mer allant jusqu'à 49 ng/g, mais la majorité des échantillons n'en contenaient pas de quantités décelables.

5. Effets sur les êtres vivants dans leur milieu naturel

On a révélé une réduction de 50% de la productivité des cultures d'algues vertes d'eau douce, en présence de concentrations de phosphate de tri-o-crésyle (TOCP) allant de 1,5 à 4,2 ng/litre, selon les espèces, les isomères méta et para étant moins toxiques. On ne dispose que de données limitées sur la toxicité aiguë du TCP vis-à-vis des invertébrés aquatiques: la CL_{50} à 48 heures pour la daphnie est de 5,6 ng/litre; la CL_{50} à 24 heures pour les nématodes est de 400 ng/litre; la dose sans effet observable à 2 semaines pour la daphnie (mortalité, croissance, reproduction) est de 0,1 mg/litre. Pour trois espèces de poissons, les valeurs de la CL_{50} à 96 heures se situaient entre 4,0 et 8700 mg/litre. Chez des truites arc-en-ciel on a constaté une mortalité d'environ 30% après exposition de quatre mois à 0,9 ng/litre de IMOL S-140 (phosphate de tri-o-crésyle à 2%) et des effets plus légers sur une période de 14 jours.

Les niveaux d'exposition au cours de ces expériences étaient beaucoup plus élevés que les concentrations susceptibles d'être rencontrées dans le milieu naturel et dans la plupart des cas, les valeurs étaient très supérieures à la solubilité des composés.

6. Cinétique et métabolisme

L'absorption, la distribution, le métabolisme et l'élimination des organophosphorés jouent un rôle déterminant dans les effets neuropathologiques retardés de ces composés.

Résumé

Chez l'homme, l'absorption percutanée du TOCP semble être au moins dix fois plus rapide que chez le chien. On observe également une importante absorption par cette voie chez le chat. L'absorption par voie orale a été observée chez le lapin. On ne possède aucune donnée de première main sur l'absorption par la voie respiratoire.

Chez le chat, on a constaté qu'après absorption, le TOCP se répartissait largement dans l'ensemble de l'organisme, la concentration la plus élevée se situant dans le nerf sciatique, qui constitue un tissu cible. De fortes concentrations de TOCP ou de ses métabolites se rencontraient également au niveau du foie, des reins et de la vésicule bilaire.

La métabolisation du TOCP s'effectue selon trois voies. La première consiste dans l'hydroxylation d'un ou plusieurs groupes méthyles et la seconde comporte la désarylation des groupements orthocrésyles. Dans la troisième, il y a encore oxydation du groupement hydroxyméthyle en aldéhyde et en acide carboxylique. L'hydroxylation constitue une étape déterminante, car le TOCP hydroxyméthylé est cyclisé pour former du phosphate cyclique d'*o*-tolyle et de saligénol, un métabolite neurotoxique relativement instable.

Le TOCP et ses métabolites sont éliminés dans les urines et les matières fécales ainsi que, en petites quantités, dans l'air expiré.

7. Effets sur les animaux d'expérience et sur les systèmes d'épreuves *in vitro*

Des trois isomères du TCP, le TOCP est de loin celui qui présente la toxicité aiguë la plus forte et qui se révèle également le plus toxique en cas d'exposition brève. Il est le seul à déterminer une neurotoxicité retardée.

Les différents paramètres toxicologiques varient beaucoup selon l'espèce (qu'il s'agisse par exemple de la mortalité aiguë ou de la neurotoxicité retardée). Le poulet est l'une des espèces les plus sensibles.

On a pu obtenir chez des espèces d'animaux de laboratoire très variées une neuropathie retardée induite par un organophosphoré (NRIOP) tant à la suite d'une exposition unique qu'à la suite d'expositions répétées. Il s'agit d'une neuropathie qui se traduit par des altérations dégénératives au niveau de la partie distale de l'axone et qui s'étend peu à peu à l'ensemble du neurone.

Elle se manifeste cliniquement par une paralysie des pattes postérieures après une période de latence caractéristique de deux à trois semaines suivant l'exposition. Une dose orale unique de 50 à 500 mg de TOCP/kg a produit une neuropathie retardée chez des poulets, des doses de 840 mg/kg ou davantage étant nécessaires pour produire une dégénérescence de la moëlle épinière chez des rats Long-Evans. C'est l'un des métabolites du TOCP, le phosphate cyclique d'o-tolyle et de saligénol, qui constitue l'agent neurotoxique actif. La sensibilité des espèces varie en sens inverse de la vitesse de métabolisation ultérieure.

On pense que la lésion biochimique qui conduit à la neuropathie consiste dans l'inhibition de l'"estérase neurotoxique". Un taux d'inhibition de plus de 65% peu de temps après une exposition au TOCP fait présager l'apparition ultérieure d'une neuropathie. La variabilité dans la réponse neurotoxique dépend également d'autres facteurs (par exemple la voie d'exposition, l'âge, le sexe, la souche). Les données disponibles ne permettent pas de définir clairement une dose sans effet observé pour cette neuropathie.

Les études de reproduction effectuées sur des rats et des souris qui recevaient des doses répétées de TOCP par voie orale, ont révélé la présence de lésions histopathologiques au niveau des testicules et des ovaires, de modifications dans la morphologie des spermatozoïdes, d'une moindre fécondité chez les deux sexes ainsi que d'une diminution de la taille des portées et de la viabilité des ratons et des souriceaux. Les données disponibles n'ont pas permis de déterminer la dose sans effet pour ce type d'anomalies imputables au TOCP. Une étude de tératogénicité effectuée sur des rats, avec des doses orales toxiques pour la mère, n'a pas donné de résultats positifs.

On n'a guère de données sur la mutagénicité et aucune sur la cancérogénicité de cette substance.

Résumé

8. Effets sur l'homme

L'ingestion accidentelle constitue la cause principale d'intoxication. Depuis la fin du dix-neuvième siècle, de nombreux cas d'intoxication dus à la contamination de boissons, de denrées alimentaires ou de produits pharmaceutiques ont été signalés. L'exposition professionnelle tient essentiellement à une absorption percutanée ou à une inhalation et certains cas d'intoxication de ce type ont été signalés. L'ingestion de préparations contaminées par du TOCP peut produire des symptômes digestifs (nausée, vomissements et diarrhées), encore que dans certains cas, c'est la polyneuropathie qui constitue le premier signe d'intoxication. Les symptômes neurologiques sont caractérisés par une période de latence. Les premiers symptômes consistent en douleurs et paresthésie aux extrémités des membres inférieurs. Il y a une légère diminution de la sensibilité cutanée et quelques fois réduction de la sensibilité vibratoire. Dans la plupart des cas, la faiblesse musculaire évolue rapidement vers une paralysie des extrémités inférieures avec ou sans extension aux membres supérieurs. Dans les cas graves, apparaissent des signes d'atteinte pyramidale. Les cas mortels sont rares mais les symptômes neurologiques peuvent être très lents à disparaître et la guérison peut prendre des mois, voire des années. L'examen histopathologique révèle une dégénérescence des axones. Les examens classiques de laboratoire ne révèlent pas d'anomalies si ce n'est parfois une augmentation de la teneur en protéines du liquide céphalorachidien. Les premiers soins consistent à réduire l'exposition en faisant vomir immédiatement le malade, dans la mesure où celui-ci est encore conscient. A plus long terme, le traitement consiste essentiellement en une réadaptation physique car on ne connaît pas d'antidote spécifique. La réaction au phosphate de tricrésyle varie considérablement d'un individu à l'autre de même que les possibilités de guérison à la suite d'une intoxication. On a signalé l'apparition de symptômes graves après ingestion de 0,15 g de TCP, alors que chez d'autres personnes, l'ingestion de quantités atteignant 1 à 2 g n'a produit aucun effet toxique. Certains malades guérissent complètement alors que d'autres présentent des séquelles marquées pendant de très longues périodes.

EVALUATION DES RISQUES POUR LA SANTE HUMAINE ET DES EFFETS SUR L'ENVIRONNEMENT

1. Evaluation des risques pour la santé humaine

On a souvent signalé des cas d'intoxication humaine par suite d'une ingestion accidentelle de phosphate de tri-*o*-crésyle ou par suite d'une exposition professionnelle. Les symptômes neurotoxiques correspondent tout d'abord à l'inhibition de la cholinestérase puis à une neuropathie retardée caractérisée par une paralysie grave.

Etant donné les variations considérables de sensibilité selon les individus, il n'est pas possible de déterminer quel est le seuil de sécurité. Des symptômes ont été signalés après ingestion de 0,15 g d'un mélange d'isomères contenant une faible proportion de TOCP; la dose minimale efficace d'isomères ortho est donc beaucoup plus faible encore. L'expérimentation animale fait également ressortir de très importantes variations selon les espèces pour ce qui concerne la réaction au TOCP et l'homme semble être particulièrement sensible.

Des cas de dermatite d'irritation et de dermatites allergiques ont été rapportés.

On peut donc considérer que l'isomère ortho et les mélanges d'isomères qui en contiennent constituent un risque de première importance pour la santé humaine.

Aucune dose n'est sans danger si elle est ingérée. Il convient de réduire au minimum l'exposition cutanée ou respiratoire.

1.1 Niveaux d'exposition

On peut considérer comme minimale l'exposition de la population générale au phosphate de tricrésyle par l'intermédiaire des divers compartiments du milieu ambiant, notamment l'eau de consommation. On a décelé du phosphate de tricrésyle à des concentrations plus fortes dans l'air urbain que dans l'air prélevé au niveau des sites de production, encore que ces valeurs soient

Evaluation

généralement faibles. Lors d'une enquête effectuée aux Etats-Unis d'Amérique on n'a pas décelé de TCP dans des échantillons de tissu adipeux humain. Il y a eu de nombreux cas d'intoxication humaine accidentelle dus à l'ingestion de médicaments, de nourriture, de farine, d'huile de cuisine et de boissons contaminés par des fluides hydrauliques ou des lubrifiants à base de TCP produits à partir d'acide crésylique. Des symptômes toxiques peuvent s'observer après ingestion de doses ne dépassant pas 0,15 g de TOCP, substance qui est présente dans le TCP produit à partir de l'acide crésylique. La contamination se produit généralement lors de la réutilisation de fûts ou de tonneaux vides qui avaient contenu un liquide hydraulique ou de l'huile lubrifiante.

1.2 Effets toxiques

L'absorption accidentelle d'une forte dose entraîne chez l'homme des troubles digestifs, à savoir une nausée d'intensité variable, pouvant aller jusqu'aux vomissements, accompagnée de douleurs abdominales et de diarrhées. En cas d'exposition à de faibles doses cumulées, une "neuropathie retardée" s'installe progressivement après une période de latence de 3 à 28 jours. Dans la plupart des cas, la faiblesse musculaire fait rapidement place à une paralysie des membres inférieurs qui peut parfois s'étendre aux mains. Dans les cas graves, on voit peu à peu appraître des signes d'atteinte pyramidale. Certaines études neurophysiologiques révèlent l'existence d'une neurotoxicité généralisée avec prolongation du temps de latence terminale et réduction relativement faible de la vitesse de conduction des nerfs moteurs. Ces constatations confirment la dégénérescence axonale qui est la principale caractéristique relevée lors des examens histopathologiques.

Le métabolite neurotoxique du TCP a été identifié; il s'agit du phosphate cyclique de o-tolyle et de saligénol qui provient lui-même des métabolites o-hydroxyméthylés. Il semble donc que la présence d'au moins un groupement o-tolyle soit nécessaire parmi les trois restes phénoliques du TCP pour qu'apparaissent des effets neurotoxiques. Le TCP produit à partir du crésol de synthèse, qui contient moins de 0,1% d'orthocrésol, n'est donc pas neurotoxique.

Les résultats d'études de toxicité subchronique effectuées sur des animaux d'expérience avec du TCP préparé à partir de crésol de synthèse, montrent que les organes cibles sont le foie et le rein; toutefois cette observation n'a pas été confirmée chez l'homme. On ne dispose pas de données suffisantes sur la mutagénicité et la cancérogénicité du TCP. On sait néanmoins que le TCP n'est pas toxique pour l'embryon de poulet.

2. Evaluation des effets sur l'environnement

Le dosage du TCP dans l'eau montre que la contamination est faible. Cet état de choses tient à la faible solubilité dans l'eau du TCP et à l'aisance avec laquelle il se décompose. Etant donnée la faible toxicité aiguë du TCP pour les organismes aquatiques, il est peu probable qu'il constitue une menace pour ces organismes.

Du fait de ses propriétés physico-chimiques, le TCP a une forte tendance à la bioaccumulation. Toutefois celle-ci ne se produit pas dans la pratique en raison de la faible concentration du TOCP dans l'environnement et les êtres vivants et de la dégradation rapide de cette substance.

Les TCP fixés aux sédiments s'accumulent dans l'environnement et on en a relevé des concentrations élevées dans les sédiments des cours d'eau et des estuaires ainsi que dans les sédiments marins. Du fait que l'on ne possède aucune information sur la biodisponibilité de ces résidus pour les organismes fouisseurs ou benthiques, ni sur les dangers qu'ils pourraient représenter, on ne peut écarter à priori la possibilité d'effets nocifs.

Il faut également mentionner la possibilité de risques localisés pour l'environnement dus à un déversement accidentel de TCP.

2.1 Niveaux d'exposition

Les TCP sont présents dans l'air, dans les eaux superficielles, dans le sol, les sédiments et les organismes aquatiques, à proximité des zones très industrialisées, encore que leurs concentrations y soient généralement faibles. En raison de la vitesse élevée de biodégradation de ces substances en milieu aqueux, il ne semble pas qu'elles puissent avoir des effets nocifs sur la faune aquatique. Il a été fait état d'une concentration extrêmement élevée en phosphates totaux de triaryle (25,55 g/kg) dans un échantillon de sol provenant d'une plantation. Cette observation montre qu'il est nécessaire de procéder à l'enfouissement des déchets.

2.2 Effets toxiques

Les algues d'eau douce sont relativement sensibles aux TCP, la concentration inhibant à 50% la croissance comprise entre 1,5 et 5,0 ml/litre. En ce qui concerne les poissons, des concentrations de TCP inférieures à 1 mg/litre (0,3-0,9 mg/litre) provoquent des signes d'intoxication chronique chez la truite arc-en-ciel, mais *Menidia notata* est plus résistant (CL_{50} de 8700 mg/litre). Les TCP n'inhibent pas l'activité cholinestérasique des poissons ou des grenouilles mais ils potentialisent l'effet des insecticides organophosphorés.

RECOMMANDATIONS

Lorsqu'on utilise des crésols tri-substitués pour la synthèse et la préparation d'autres composés, il est préférable d'utiliser des isomères para et méta purifiés afin d'éviter toute synthèse accidentelle de dérivés orthosubstitués.

RESUMEN

1. Identidad, propiedades físicas y químicas, y métodos analíticos

El fosfato de tricresilo (FTC) es un líquido ininflamable, no explosivo, incoloro y viscoso. Su coeficiente de partición entre el octanol y el agua (log P_{ow}) es de 5,1. Se hidroliza con facilidad en un medio alcalino dando fosfato de dicresilo y cresol, pero es estable en medios neutros y ácidos a temperaturas normales.

El método analítico de elección es la cromatografía de gases con un detector sensible al nitrógeno-fósforo o un detector fotométrico de llama. El límite de detección en una muestra de agua es de 1 ng/litro aproximadamente. El FTC se extrae con facilidad de las soluciones acuosas con distintos disolventes orgánicos. Se utiliza habitualmente para la extracción la cromatografía en columna de florisil, pero es difícil separar el FTC de los lípidos con este método. Se han recomendado para esa finalidad otros métodos de extracción (GPC, cromatografía en carbón vegetal activado y Sep-pak C-18). Los reactivos analíticos están contaminados a menudo con cantidades infinitesimales de FTC debido a su amplio uso. Por consiguiente, la obtención de datos fiables en el análisis de cantidades infinitesimales de FTC requiere un procedimiento cuidadoso.

2. Fuentes de exposición humana y ambiental

El FTC se produce habitualmente por reacción de cresoles con oxicloruro de fósforo. Existen dos fuentes industriales de cresoles: el "ácido cresílico", obtenido como residuo de los hornos de carbón de coque y del refino del petróleo; y los "cresoles sintéticos", preparados a partir del cimeno por oxidación y degradación. Como resultado, el FTC es una mezcla de varios fosfatos triarílicos.

El FTC se utiliza como plastificante en los plásticos vinílicos, y también como pirorretardante, aditivo para lubricantes de presión extrema y líquido ininflamable en los sistemas hidráulicos.

3. Transporte, distribución y transformación en el medio ambiente

El paso de FTC al medio ambiente se debe principalmente a su uso final, pues la liberación en el curso de la fabricación es escasa. En 1977 se calculó que el paso total al medio ambiente en los Estados Unidos de América fue de 32 800 toneladas.

Debido a su escasa hidrosolubilidad y a su elevada adsorción por los materiales en partículas, el FTC se absorbe con rapidez en los sedimentos de ríos o lagos y en el suelo. Su biodegradación en el medio acuático es rápida, quedando casi terminada en el agua de río en cinco días. El isómero orto se degrada con una rapidez ligeramente mayor que los isómeros meta o para. El FTC se biodegrada con facilidad en el fango de los alcantarillados, presentando una semivida de 7,5 horas; la degradación en 24 horas alcanza el 99%. La degradación abiótica es más lenta, dando una semivida de 96 días.

Se midieron los factores de bioconcentración de 165-2768 en varias especies de peces en el laboratorio utilizando FTC radiomarcado. La radiactividad desapareció rápidamente al cesar la exposición, observándose semividas de depuración comprendidas entre 25,8 y 90 horas.

4. Niveles medioambientales y exposición humana

En el Japón se han medido concentraciones atmosféricas de FTC de hasta 70 ng/m^3, pero alcanzaron un máximo de sólo 2 ng/m^3 en una instalación de fabricación de los Estados Unidos de América. En este país, el aire del medio laboral contenía menos de 0,8 mg/m^3 en una nave de llenado de barriles de aceite lubricante y 0,15 mg/m^3 (fosfatos totales) en una planta de troquelado de zinc para automóviles. En el Canadá, las concentraciones de FTC medidas en el agua potable fueron bajas (0,4 a 4,3 ng/litro) y el producto resultó indetectable en el agua de pozo. Las concentraciones observadas en las aguas de ríos y lagos son con frecuencia apreciablemente mayores. Sin embargo, ello se debe a la presencia de sedimentos en suspensión en los que queda fuertemente absorbido el FTC.

Las concentraciones en los sedimentos son altas en los ríos y en el mar, habiéndose observado valores de hasta 1300 ng/g y 2160 ng/g, respectivamente.

Se observaron concentraciones altas en el suelo y la vegetación en los alrededores de instalaciones de fabricación.

Se han señalado restos en peces y mariscos de hasta 40 ng/g, pero la mayor parte de los animales examinados no contenían residuos apreciables.

5. Efectos sobre los seres vivos del medio ambiente

La productividad primaria de cultivos de algas verdes de agua dulce quedó reducida en el 50% mediante la adición de fosfato de tri-o-cresilo a razón de 1,5 a 4,2 ng/litro, según las especies, mientras que los isómeros meta y para resultaban menos tóxicos. Son limitados los datos referentes a la toxicidad aguda del FTC para los invertebrados acuáticos: la CL_{50} en 48 horas para Daphnia es de 5,6 ng/litro y la CL_{50} en 24 horas para los nematodos es de 400 ng/litro; la concentración NOEL (mortalidad, crecimiento, reproducción) en dos semanas para Daphnia es de 0,1 mg/litro. Los valores de CL_{50} en 96 horas para tres especies de peces se hallaban comprendidos entre 4,0 y 8700 mg/litro. La trucha irisada presentó una mortalidad del 30% aproximadamente después de una exposición de cuatro meses a una concentración de 0,9 ng/litro de IMOL S-140 (fosfato de tri-o-cresilo al 2%) y efectos menores en un periodo de 14 días.

Los niveles de exposición utilizados en esos experimentos fueron mucho mayores que las concentraciones que probablemente pueden hallarse en el agua en el medio ambiente y, en la mayoría de los casos, excedían en gran manera a la solubilidad de los productos.

6. Cinética y metabolismo

La absorción, distribución, metabolismo y eliminación de los organofosfatos son elementos críticos en los efectos neuropáticos tardíos de estos productos.

La absorción cutánea del FTOC en el hombre parece ser por lo menos de un orden de magnitud más rápida que en los perros. También se observa una absorción cutánea significativa en el gato. En el conejo se ha señalado la absorción oral del producto. No hay información directa sobre la absorción por inhalación.

En estudios efectuados en gatos se observó que el FTOC absorbido se distribuye ampliamente por todo el organismo, hallándose la máxima concentración en el nervio ciático, que es el tejido diana. Otros órganos en los que se encuentran altas concentraciones de FTOC y de sus metabolitos son el hígado, los riñones y la vesícula biliar.

El metabolismo del FTOC sigue tres vías. La primera es la hidroxidación de uno o más grupos metílicos y la segunda es la desarilación de los grupos o-cresilo. La tercera vía es la oxidación ulterior del grupo hidroximetilo para dar aldehído y ácido carboxílico. La etapa de hidroxilación es decisiva porque el FTOC hidroximetílico forma un producto cíclico, el fosfato cíclico de o-tolilo saligenina, metabolito neurotóxico relativamente inestable.

El FOTC y sus metabolitos se eliminan por la orina en las heces, junto con pequeñas cantidades en el aire espirado.

7. Efectos en los animales de experimentación y en sistemas de prueba *in vitro*

Entre los tres isómeros del FTC, el FOTC es con gran diferencia el más tóxico en la exposición aguda y a corto plazo. Es el único isómero que produce neurotoxicidad tardía.

Existe una amplia variabilidad entre especies en lo que respecta a los distintos puntos finales tóxicos (por ejemplo, letalidad aguda, neurotoxicidad tardía) de la exposición al FOTC, siendo el pollo una de las especies más sensibles.

Se ha producido neuropatía tardía inducida por organofosfatos mediante la exposición única y repetida en una amplia gama de animales de experimentación; este trastorno se clasifica como una "neuropatía de muerte sobre el dorso". Se producen lesiones degenerativas en el axón distal, que se extienden con el tiempo hacia el cuerpo de la célula.

Resumen

Los signos clínicos consisten en la parálisis de las patas traseras después de un intervalo característico de 2-3 semanas a partir de la exposición. Una sola dosis oral de 50-500 mg de FOTC/kg produjo la neuropatía tardía en pollos, mientras que se necesitaron dosis de 840 mg/kg o más para producir la degeneración de la médula espinal en ratas Long-Evans. El metabolito fosfato cíclico de o-tolilo saligenina es el agente neurotóxico activo. La sensibilidad de las especies guarda correlación inversa con la tasa de metabolismo ulterior.

Se cree que la inhibición de la "esterasa de la neurotoxicidad" es la lesión bioquímica que conduce a la neuropatía tardía inducida por organofosfatos; la inhibición de más del 65% poco después de la exposición al FOTC permite prever una neuropatía ulterior. En la variabilidad de la respuesta neurotóxica al FOTC influyen factores distintos del metabolismo (por ejemplo, vías de exposición, edad, sexo, estirpe). Los datos disponibles no permiten establecer un nivel neto sin efectos observados para la neuropatía tardía.

Los estudios de reproducción en ratas y ratones sometidos a una exposición oral repetida al FOTC mostraron lesiones histopatológicas de los testículos y los ovarios, alteraciones morfológicas del esperma, disminución de la fecundidad en ambos sexos, y descenso del tamaño y viabilidad de las crías. Los datos disponibles no permiten establecer un nivel neto sin efectos en lo que respecta a la acción del FOTC sobre la reproducción. Un estudio de teratogenicidad en ratas, utilizando dosis orales que producían toxicidad materna, dio resultados negativos.

Se dispone de escasa información sobre la mutagenicidad y de ninguna sobre la cancerogenicidad.

8. Efectos en la especie humana

La ingestión accidental es la principal causa de intoxicación. Desde fines del siglo XIX se han observado numerosos casos de intoxicación por contaminación de bebidas, alimentos o medicamentos. La exposición profesional se produce principalmente por absorción cutánea o inhalación, habiéndose registrado algunos casos de envenenamiento. La ingestión de preparaciones que contienen FOTC puede ir seguida de síntomas gastrointestinales (náuseas, vómitos y diarrea), aunque en algunos casos la polineuropatía es el primer signo de intoxicación. Los síntomas neurológicos suelen ser tardíos. Los síntomas iniciales consisten en dolor y parestesia de las extremidades inferiores. Puede observarse una alteración moderada de las sensaciones cutáneas y a veces del sentido de la vibración. En la mayoría de los casos, la debilidad muscular evoluciona rápidamente hasta dar una marcada parálisis de las extremidades inferiores, con o sin participación de las superiores. En los casos graves aparecen signos piramidales. Son raras las defunciones, pero la recuperación de los síntomas y signos neurológicos puede ser extremadamente lenta y durar varios meses o años. El examen histopatológico muestra la presencia de degeneración del axón. Los análisis corrientes de laboratorio no muestran hallazgos anormales, pero cabe observar un aumento de la concentración de las proteínas en el líquido cefalorraquídeo. Los primeros auxilios consisten en la reducción de la exposición provocando el vómito inmediatamente después de la ingestión, siempre que el paciente esté consciente. El tratamiento fundamental a largo plazo consiste en la rehabilitación física, no conociéndose ningún antídoto específico. Existen grandes variaciones entre las personas en la respuesta al FTC y en la recuperación después de producirse los efectos tóxicos. Se han registrado síntomas graves después de la ingestión de 0,15 g de FTC, mientras que otras personas no presentaron efecto tóxico alguno después de ingerir 1-2 g. Ciertos enfermos se recuperan por completo, mientras que otros conservan efectos marcados durante un periodo apreciable.

EVALUACION DE LOS RIESGOS PARA LA SALUD HUMANA Y DE LOS EFECTOS EN EL MEDIO AMBIENTE

1. Evaluación de los riesgos para la salud humana

Se han registrado con frecuencia intoxicaciones humanas por ingestión accidental de fosfato de tri-*o*-cresilo (FTOC) o por exposición laboral de trabajadores. La vía probable de la exposición laboral es la absorción cutánea. Entre los síntomas neurotóxicos figuran la inhibición inicial de las colinesterasas y la neuropatía tardía ulterior caracterizada por parálisis grave.

Debido a la considerable variación existente en la sensibilidad de las personas al FTOC, no puede establecerse un nivel de exposición sin riesgo. Se han señalado síntomas por ingestión de 0,15 g de una mezcla de isómeros con baja proporción de FTOC; así pues, la dosis efectiva mínima del ortoisómero es muy inferior. Los estudios efectuados en animales muestran considerables variaciones entre especies en la respuesta al FTOC, y las personas parecen ser especialmente sensibles.

Se han señalado casos de dermatitis irritante y alérgica.

En consecuencia, tanto el ortoisómero puro como las mezclas de isómeros que contienen FTOC se consideran riesgos importantes para la salud humana.

No existe un nivel inocuo de ingestión. La exposición al producto por contacto cutáneo o inhalación debe reducirse al mínimo.

1.1 Niveles de exposición

Puede considerarse mínima la exposición de la población general al fosfato de tricresilo (FTC) por conducto de distintos medios ambientales, incluida el agua de beber. Se han observado concentraciones de FTC relativamente más altas en el aire urbano que en el aire recogido en los emplazamientos de fabricación, aunque los niveles suelen ser bajos. En un estudio efectuado en los Estados

Unidos de América no se detectó el FTC en muestras de tejido adiposo humano. Se han observado numerosos casos de intoxicación humana accidental por ingestión de medicamentos, alimentos tales como harina y aceite de cocinar, y bebidas contaminados con líquido hidráulico o aceite lubricante que contenían FTC producido a partir del "ácido cresílico". Pueden observarse síntomas tóxicos después de la ingestión de sólo 0,15 g de fosfato de tri-o-cresilo, componente del FTC obtenido a partir de ácido cresílico. El origen habitual de la contaminación ha consistido en la reutilización de barriles o bidones vacíos utilizados previamente para contener líquido hidráulico o aceite lubricante.

1.2 Efectos tóxicos

La exposición humana accidental a una sola dosis alta ocasiona trastornos gastrointestinales que varían de las náuseas ligeras a las intensas con vómitos, dolor abdominal y diarrea. En el caso de la exposición a pequeñas dosis acumuladas, aparece progresivamente la "neurotoxicidad tardía" después de un periodo latente de 3-28 días. En la mayor parte de los casos, la debilidad muscular pasa rápidamente a ser una marcada parálisis de las extremidades inferiores, con o sin afectación de las manos. En los casos graves aparecen progresivamente signos piramidales. Algunos estudios neurofisiológicos muestran fenómenos extendidos de neurotoxicidad y prolongación de las latencias terminales, con disminución relativamente pequeña de la velocidad de conducción de los nervios motores. Ello confirma los signos de degeneración del axón, que es la característica principal observada en los exámenes histopatológicos.

Se ha identificado el metabolito neurotóxico del FTC como el fosfato cíclico de o-tolilo saligenina, derivado de los metabolitos o-hidroximetílicos. Parece pues que los efectos neurotóxicos exigen la presencia por lo menos de un grupo o-tolilo entre las tres porciones fenólicas del FTC. Ello significa que no es neurotóxico el FTC producido a partir de cresol sintético, que contiene menos del 0,1% de o-cresol.

Los estudios de toxicidad subcrónica efectuados en animales con FTC obtenido de cresol sintético muestran que

Evaluación

los órganos diana son el hígado y los riñones, pero esa observación no se ha confirmado en los casos de intoxicación humana. No se dispone de datos apropiados sobre la mutagenicidad y la cancerogenicidad. El FTC no es tóxico para los embriones de pollo.

2. Evaluación de los efectos en el medio ambiente

La medición de las concentraciones de FTC en el agua ambiental ha mostrado que existen sólo niveles bajos de contaminación. Ese hecho refleja la escasa hidrosolubilidad y la fácil degradabilidad del producto. Dado que la toxicidad aguda del FTC para los seres acuáticos es también baja, es improbable que represente una amenaza para ellos.

Debido a las propiedades fisicoquímicas del FTC existen altas posibilidades de bioacumulación. Sin embargo, ello no se produce en la práctica porque las concentraciones de FOTC en el medio ambiente y en los organismos vivos son bajas y porque los productos se degradan con rapidez.

El FTC fijado por los sedimentos se acumula en el medio ambiente, habiéndose medido concentraciones altas en los sedimentos de ríos, estuarios y mares. Dado que se carece de información sobre la biodisponibilidad de esos residuos para los seres vivos que se hallan en madrigueras o en el fondo o sobre sus riesgos, no puede descartarse la posibilidad de que aparezcan efectos en tales especies.

La fuga de FTC suscita riesgos para el medio ambiente local.

2.1 Niveles de exposición

El FTC se halla en el aire, las aguas de superficie, el suelo, los sedimentos y los seres vivos acuáticos cerca de las zonas muy industrializadas, aunque en concentraciones habitualmente bajas. Debido a la alta tasa de biodegradación del FTC en el medio acuoso, no se considera que afecta adversamente a los seres vivos acuáticos. En un estudio se encontró una concentración extremadamente alta de fosfatos triarílicos totales (26,55 g/kg) en una muestra de suelo obtenida en el patio de una planta de producción. Ello sugiere la necesidad de eliminar los residuos por relleno de terrenos.

2.2 Efectos tóxicos

Las algas de agua dulce son relativamente sensibles al FTC, cuya concentración inhibidora del 50% del crecimiento es de 1,5 a 5,0 mg/litro. Entre las especies de peces, la trucha irisada sufre el efecto de concentraciones de FTC inferiores a 1 mg/litro (0,3-0,9 mg/litro), con signos de intoxicación crónica, pero el pez argentado de las mareas es más resistente (CL_{50} de 8700 mg/litro). El FTC no inhibe la actividad colinesterásica seca de peces y ranas, pero tiene un efecto sinérgico con la actividad insecticida de los órganos fosforados.

RECOMENDACIONES

Cuando se utilizan cresoles trisustituidos en la síntesis y fabricación de otros productos, es necesario emplear isómeros meta y para purificados para evitar la síntesis accidental de productos orto-sustituidos.

www.ingramcontent.com/pod-product-compliance
Ingram Content Group UK Ltd.
Pitfield, Milton Keynes, MK11 3LW, UK
UKHW021310180426
11947UKWH00015B/1131